Nature's Unruly Mob

Nature's Unruly Mob

Farming and the Crisis in Rural Culture

PAUL GILK

WIPF & STOCK · Eugene, Oregon

NATURE'S UNRULY MOB
Farming and the Crisis in Rural Culture

Wipf & Stock
A Division of Wipf and Stock Publishers
199 W. 8th Ave., Suite 3
Eugene, OR 97401
www.wipfandstock.com

ISBN 13: 978-1-60608-737-4

Manufactured in the U.S.A.

For my friend Maynard Kaufman,
whose back-to-the-land life is an example for us all

I finally realized that space travel is not new: it is only the final stage of a departure process that actually began long ago. Our society really "left home" when we placed boundaries between ourselves and the earth, when we moved en masse inside totally artificial, reconstructed, "mediated" worlds—huge concrete cities and suburbs—and we aggressively ripped up and redesigned the natural world. By now, nature has literally receded from our view and diminished in size. We have lost contact with our roots. As a culture, we don't know where we came from; we're not aware we are part of something larger than ourselves. Nor can we easily find places that reveal natural processes still at work.

This is exacerbated for Americans in particular, since our country is made up almost entirely of immigrants whose original connections with a homeland were severed, and who have no special attachments to the soil we live on. The Native Americans, who do have roots here, are not nearly as enthusiastic about leaving the earth as the rest of us are . . . .

—Jerry Mander,
In the Absence of the Sacred, p. 148

Since enhancing the prestige of farming as an occupation is critical to developing the sun-based regional agriculture we need, the White House should appoint, in addition to a White House chef, a White House farmer. This new post would be charged with implementing what could turn out to be your most symbolically resonant step in building a new American food culture. And that is this: tear out five prime south-facing acres of the White House lawn and plant in their place an organic fruit and vegetable garden.

—Michael Pollan,
"Farmer in Chief,"
in *The New York Times Magazine* (October 12, 2008), p. 71

Contents

Acknowledgments

I<small>N THE FALL OF</small> 1985, Jack Miller and Pauline Redmond of Anvil Press established a Woodbox Fund that enabled Jo Wood and me to live several months with them and their young son Daniel. We helped produce a couple issues of Jack's regional magazine, *North Country Anvil*. From that collaboration came an invitation to prepare *Nature's Unruly Mob* for publication as a special issue of the magazine.

I not only want to thank all those who helped with the initial effort—Jo, Pauline, Jack, Rhoda Gilman, Maynard Kaufman, Cherie Lozier, Vic Ormsby, Jim Mullen, John Grobner, and Joanne Klees (Pauline and Jim are no longer with us)—but especially now I thank Jack for so graciously allowing Wipf and Stock to republish the *Mob* without hesitation or reservation.

I also want to thank Ted Lewis for his trust in the project, Helena Norberg-Hodge for her enthusiasm, Howard Zinn for his amazing graciousness, and Carol Ann Okite for her cheerful, steady, competent keyboard work, without which there would simply be no *Mob* to attack the fortified Best Seller List with peasant pitchforks and hoes.

Foreword

BOOKS LIKE THIS ONE—BOOKS that remind us of our deep and abiding connection to one another and to the earth—are rare in today's high-tech, fast-paced consumer culture. Drawing on both extensive readings and first-hand experience and observation, Paul Gilk explores the ways in which industrial society can become healthier and more life-sustaining, weaving a compelling argument for the revitalization and expansion of rural culture. His insights and sentiments in this regard are reminiscent of those of the great philosopher and writer Wendell Berry.

For many urban dwellers, envisioning a rural renaissance can seem unrealistic, even undesirable. This is understandable: our media and—even more importantly, as Paul points out—our system of education, have been promoting an urbanizing model of economic development as "progress" for many generations now. As part of this process, small farmers all over the world have been and continue to be marginalized and disempowered. When modern, urban people experience rural life, it is therefore usually through contact with those who have been underpaid and undervalued, whose entire way of life has been undermined.

I am very sympathetic to Paul's thinking because I have had the rare privilege of living and working in a rich and thriving land-based culture, in which small farmers are valued and respected as the source of the most essential goods of all: the food everyone needs to survive. Over three decades ago, when Paul was returning to rural life in his native Wisconsin, I went to Ladakh, a preindustrial culture high on the Tibetan plateau. I was a linguist and was there to learn Ladakhi. I soon discovered that this ancient culture had far more to teach me and the outside world than just its language.

The vast majority of Ladakhis were self-supporting farmers, living in small, scattered settlements in the high desert of the Indian Himalayas. Though natural resources were scarce and hard to obtain, the Ladakhis had a remarkably high standard of living—producing not just their basic needs, but beautiful art, architecture, and jewelry. Because the culture fulfilled fundamental human needs while respecting natural limits, there were no signs of environmental stress. The way of life was rich in other ways as well: people worked at a gentle pace and enjoyed a degree of leisure unknown to most people in the West. The various connecting relationships in the traditional system were mutually reinforcing, encouraging harmony and stability. In fact, the Ladakhis were the most vital, happy, and contented people I have ever known. It became clear to me that this traditional nature-based society was far more sustainable, both socially and environmentally, than the Western consumer society I had been living in.

Shortly after I arrived, however, Ladakh was thrown open to economic development. Over the next decades, I witnessed a process of change that, in many parts of the world, has taken hundreds of years. The area was flooded with imported goods, including subsidized and processed food. Advertising and mainstream media, tourists, so-called "modern" education, and other trappings of conventional development descended on the Ladakhis like an avalanche. The impacts included unemployment (which was previously nonexistent), a widening gap between rich and poor, and perhaps most strikingly, cultural self-rejection. The young, in particular, started seeing their own way of life as backward and inferior compared to the romanticized images of an urban, consumer lifestyle.

Somewhere early on during that painful transition, I read E. F. Schumacher's *Small is Beautiful: Economics as if People Mattered.* The book strengthened my convictions and started me on a journey of working with the Ladakhis to find ways of meeting the modern world that would not undermine their local culture and economy. I started a small organization, The Ladakh Project, which has since grown into the International Society for Ecology and Culture (ISEC).

Until the last year or so, I'd never heard of Paul Gilk. As it turns out, Paul's Swiss wife, Susanna, read my book about tradition and change in Ladakh, *Ancient Futures: Learning from Ladakh*, and urged Paul to read it, too. He then sent a warm and friendly letter to the ISEC office. Later he sent me a copy of *Nature's Unruly Mob: Farming and the Crisis in Rural Culture.*

What Paul says regarding redemption of the past is absolutely essential for our survival. He understands the vital importance of small-scale, diversified agriculture—having grown up on a first-generation homestead farm in northern Wisconsin, about a hundred miles south of Lake

Superior. This was a place with rocky, glacial soil and unpredictable frosts, with workhorses, and nearly surrounded by cut-over forest.

We are almost exactly the same age. When I first went to Ladakh, Paul was returning to the family farm after nearly ten years in inner-city St. Louis. In 1976, he built what he calls a "shack" in the woods and lived there, without electricity or running water, for most of the next twenty years. He did occasional farm work, woods work and carpentry; kept a garden; burned wood for heating and cooking; studied and wrote. He lived a life, he says laughingly, of "voluntary poverty," a kind of poverty reminiscent of Thoreau at Walden Pond.

While in St. Louis, Paul says, he not only began to miss rural life, but was increasingly aware that small farms were dying. He asked people he considered informed and intelligent to explain to him why this was happening. The answers were trite and totally unsatisfactory, so he began to read history. Getting to the bottom of this mystery became a passion. Grounded from childhood in a life of modest subsistence, yet thoroughly drenched through schooling in notions of technological progress, Paul spiraled back to his roots—fortified with an intellectual determination to discover why there was such an incredible bias against rural life.

A lot of thinkers have been influenced by the idea that the transition from hunter-gatherer culture to early farming was the primal disaster for nature and humanity—that virtually all of our modern woes can be traced back to this "Fall." As part of this view, many environmentalists support maximized urbanization as a way, or at least as a preliminary step, toward freeing the countryside from human habitation to the fullest possible extent. In theory, extremely concentrated cities would then be balanced by a vast natural wilderness.

Paul's understanding is very different. He sees the development of early farming as a natural progression, proceeding from the diligence of gatherers whose attentiveness toward plants gave rise to gardening and horticulture. Crisis did occur, but not from small-scale farming. Crisis came with predatory cities, and with autocratic leadership that utilized militarism and slavery to extract wealth from others. This exploitative, expansionist system has grown into the global economy of today.

We live in a time of great change and uncertainty about the future. More and more people recognize that continuing on our path of uncontrolled economic growth is little short of madness. Paul points out that "Despite a growing uneasiness among many of us regarding pollution, environmental damage, urban sprawl, an underpopulated countryside, and military intervention on behalf of an economic ideology that falsely parades itself as *conserve*-ative, the official commitment to industrial growth, and to the required energy consumption by which to promote that growth, remains unchanged." He offers a thorough analysis of how the entire structural system—education, research and technology, government policy and subsidies, trade agreements, international aid, etc.—ignores human needs and natural limits; it is an economic system that is blind to its own blindness. And it is a system that is bringing us closer every day to environmental, economic, and social suicide.

It is as clear in New York, Melbourne, and rural Wisconsin as it is in Ladakh: the consumer culture, based on competition and individualism, has not brought us greater happiness. In fact, depression and disorders resulting from low self-esteem are at an all-time high around the world. The tremendous pressures in the modern economy rob us of time with our families and other loved ones. Paul's call for the rebuilding of real culture and community—a necessary foundation for a healthy sense of identity—is fortunately beginning to be answered in many community-building initiatives around the world. In addition, many individuals and groups are working to take back from corporations the control they exert over our economies and to increase local and regional economic autonomy. Paul writes with passion and experience about the importance of these efforts in bringing about a more positive future. He also provides us with valuable perspectives on the roots of these movements, particularly the Luddites and anarchists, who have been gravely misrepresented in our industrialized culture.

Paul rightly puts agriculture at the center of all these movements: growing, harvesting, processing, and eating food connects us to the earth and to each other like no other activity. Encouraging localized food systems would stem the tide of urbanization and ensure greater food security, along with a whole host of other economic, social, and environmental benefits. Farmers would fare better financially and have an incentive to diversify, which has invaluable ecological benefits. As consumers, we would enjoy healthier food and lower prices. Efforts to reconnect people to the land are more important today than ever before, and local food economies are the most strategic way of doing so.

The enormous privilege I have had of living in Ladakh has shown me that in order for us to have genuine progress, we cannot ignore the past. Paul Gilk, through his own experience, has come to the same under-standing. Ancient ways of life teach us about living closely and harmoni-ously with the land, meeting our basic needs without compromising our health—mental and physical—or that of the natural environment. Paul shows us that by "slowing down" we can build human-scale economies once again: liveable, durable, holistic cultures that are allied closely with nature and natural processes. I share his vision of a world where we em-brace our diverse cultural, biological, and agricultural heritages and, with this invaluable knowledge, move towards a healthier and happier future.

— Helena Norberg-Hodge

Introduction

I T IS WITH WARM affection that I look back on the publication of *Nature's Unruly Mob* by Anvil Press in 1986. There was, at the time, yet another episode of the perennial "farm crisis" working its way through the farm community (referenced in "the 40 cents per hundredweight assessment on milk for dairy herd buyout" in chapter 13), and I had a certain rosy hope the *Mob* would be useful in the current agitation. (Intellectuals are prone to such fanciful indulgences.) The Anvil shop was an occasional hub for farm activists.

Anvil Press was installed in a glorified garage, right across the dead-end alley from an old, two-story, clapboard house that Jack Miller and his wife Pauline Redmond rented from a relative of Jack's, all of it nestled within the village of Millville, a tiny farm town along the Zumbro River in southeastern Minnesota, a river with limestone bluffs rising haphazardly from either bank, with pockets of thick woods on hillsides, in pastures, and in hollows.

It was a picturesque scene in its entirety—the river, the bluffs, the village, the shop with woodsmoke drifting from the chimney, Pauline and Jack with their pipes, and, in the shop, the random scattering of cups and mugs, half-full of cold, bitter tea, with the string from the tea bag wrapped precisely around each mug's handle. The printing equipment was substantial but obsolete, all of it miraculously kept running by a charming hermit named Harvey Melcher, who had an alcohol problem and who died, eventually, by running his car into a statue of the Virgin Mary. *North Country Anvil* was a magazine of eccentricity from top to bottom, from beginning to end, most of it endearing and some of it tragic, in a larger world of stifling standardization and oppressive cliché.

Jack had been an Associated Press reporter based in Washington, D.C. Two stories from his reporter days stick in my mind, both illustrative of why he quit the news business. One had to do with machine guns installed around the White House, in the riotous aftermath of Martin Luther

King's assassination—a story Jack wrote but his editors chose not to print. The other story concerned extensive interviews with sailors who had been aboard ships involved in the infamous Gulf of Tonkin "incident," the "incident" that enabled Lyndon Johnson to get greater war powers through the Senate. Those interviews, conducted in collaboration with a colleague, fully discredited the Tonkin affair; but, at the editorial level, the story was rewritten into innocuousness. Jack was disgusted. He was *so* disgusted he decided to start a regional quarterly magazine, a decision that resulted in *North Country Anvil*, fully devoted to "hammering it out."[1]

It was in the early 1980s when I first learned of the magazine and, appropriately, I heard about it from a small, bent, bachelor sheep farmer, retired, with a shaggy, untrimmed beard and twinkly blue eyes, whose name was Andreas Odberg. How Andy had learned about the magazine was unclear, and I never did figure it out.

Within a short time, I was submitting articles to the *Anvil*. In 1984, with my (then) wife Jo Wood, I went for the first time to Millville, crossing the sprawling Mississippi at Wabasha, quickly became friends with both Jack and Pauline, and got a little volume of poetry printed in their shop. Jo and I stayed with Jack and Pauline, helped cook and bake, did childcare, and cut firewood.

We were invited back in 1985 to guest-edit a special Green issue of the magazine, and that led to talk of publishing *Nature's Unruly Mob*. It fell to me to typeset the manuscript, so Jack loaned me an electric typewriter to take home. I put that typewriter on the tailgate of my father's pickup, in his garage back in northern Wisconsin, and had, I suppose it could be said, a tailgate party for the *Mob*. (Jo and I lived in a shack in the woods and had neither electricity nor running water.) So *Nature's Unruly Mob* was originally published by Anvil Press, now defunct but fondly remembered.

One of Jack's sustained and sustaining themes in *North Country Anvil* was the plight of small-scale farming, the door through which *Nature's Unruly Mob* slipped into print, initially as a special issue of the magazine, also in 1986. For a while, I imagined we could sell the *Mob* for a "Jefferson"—that is, a two-dollar bill, a form of federal currency still in circulation but rarely used. Besides being utterly unrealistic economically, this fanciful idea (which still tickles me) met the convergence of

---

1. For a fuller account of the life of *North Country Anvil*, see Rhoda Gilman's labor of love, *Ringing in the Wilderness*, published by Holy Cow! Press in 1996. "Hammering It Out" was Jack's quarterly editorial.

harsh fact and troubling metaphor: the public wasn't much interested in rural culture, and the circulation of two-dollar "Jeffersons," never great to begin with, began to be phased out by the U.S. Treasury. I took this as yet another indication—gobs of Alexander Hamilton ten-dollar bills, almost no Thomas Jefferson two-dollar bills—of how Hamiltonian industrialism had beaten Jeffersonian agrarianism into oblivion.

Nature's Unruly Mob was my initial effort (in a book-length format) to explore how and why civilization had hijacked agriculture. But it was a very difficult intellectual and a very painful spiritual process by which I had come to this understanding. Not that the facts were all that hard to grasp. The facts were as plain as the noses on our faces. What was difficult to slip out of was the cultural convention that civilization is God's arm, God's presence on Earth, that civilization is under sacred guidance and divine protection, and that the famous "Unseen Hand," invisibly directing economic arrangements in our civilized system, is, *ipso facto*, God's hand, probably without any dirt under the fingernails.

Virtually all of us were raised with the Augustinian teaching that God, as The Ultimate Being in Control, gives and sustains kingdoms, empires, and civilizations. Although we may not have been explicitly taught that doctrine by chapter and verse (though it's easily found in *The City of God*, either in Book IV:33 or Book V:1), the Augustinian assertion has simply saturated our cultural sensibility. It's the water our mental fish swim in. To have traced the history of agriculture, therefore, and to have seen that civilization emerged from the armed theft—the expropriation—of agricultural abundance, was to confront ethically and morally the pious Augustinian creed regarding the godly origins of kingdoms and empires. That is, if God gives and sustains civilization, that sustaining gift arrived with theft, slavery, and institutionalized militarism. The facts regarding this early civilized impoundment of agriculture—the impoundment of both its production *and* its culture—provoked in me a religious crisis: if the historical facts were true, then either God is a cruel and cunning aristocrat or civilization does not have divine sanction. I could not believe the former and therefore had to believe the latter. It is almost impossible to express what an extraordinary spiritual struggle and yet liberating political revelation this was for a farm boy raised with quasi-fundamentalist convictions.

But there were more insights to come. One such insight was that our virtual worship of civilization, as a concept of transcendent cultural goodness and overarching spiritual superiority (American Exceptionalism is largely built of this stuff), keeps us collectively in the grip of an enormously destructive, pseudodivine illusion. The illusion is that civilization is fundamentally and eternally good, a gift from God (for the conventionally religious) or (for the conventionally irreligious) the supreme attainment of human nobility on Earth, something far, far above any or all noncivilized forms of culture. As far as I can tell, this is exactly why we are so politically incapacitated and spiritually quiescent as truly massive crises arise around us—crises whose roots lie in what Lewis Mumford calls civilization's "traumatic institutions." [II] How can we respond to civilization's accrued and accelerating negative impacts when civilization is a metaphysical construct containing no negatives?

Life on Earth no doubt faces great danger from many sources and directions—disease epidemics, earthquakes, tidal waves, hurricanes, collisions with meteors, and so forth. But no dangers are greater or more corruptingly deadly than those contained within and magnified by civilization itself. Various kinds of deadly toxins, induced extinctions, climate change, massive cultural breakdowns, burgeoning human populations, apocalyptic weaponry, global surveillance, and endless proliferation of crises related to "terrorism"—all derive from and are outgrowths of the brutal imposition of civilization on evolved nature and noncivilized folk cultures, the entire process hugely intensified since the fifteenth century and transformed into globalized mass imperialism by the Industrial Revolution. Two world wars later, America boasts being the only Superpower left standing on Earth, a solo stature it openly intends to maintain and extend. "Globalization" now implies the imposition of industrialized civilization into every nook and cranny in the world, replete with angry resistances, especially from Islamic people, even as the enlarging standard of living, American style, distresses global ecology with ever greater disturbances. This is our present moment of steadily growing crisis.

Breaking the shell of our intellectual and spiritual captivity to the doctrine of civilization is therefore an unconditional precondition for ecological restoration and cultural revival. Nothing is more politically

---

II. Trauma, says Erik Erikson on page 98 of *Gandhi's Truth*, means "an experience characterized by impressions so sudden, or so powerful, or strange that they cannot be assimilated at the time and, therefore, persist from stage to stage as a foreign body seeking outlet or absorption and imposing on all development a certain irritation causing stereotypy and repetitiveness." Erikson, of course, as a trained psychoanalyst, was talking about developmental stages in a psychological sense; but I see no reason why such description would not apply to history.

urgent than waking from our civilized somnolence, from our image of God as cruel aristocrat, and from our smug presumption that civilization is always blessed and fully righteous. The following chapters contain, I hope, the dawn of such an awakening.

—Paul Gilk

1

Organic Intelligence

AS BUSINESS, INDUSTRY, TECHNOLOGY, and science have expanded in this country, rural culture has contracted both in quality and in bulk. And, as the westward progression of white settlements proliferated and accelerated, many new communities were formed with little or no reference to the lay of the land, to walking trails or footpaths, or even to roadways created for the ease of horse-drawn vehicles. Not only were roads laid out by straight line and geometric grid, whole towns and even cities were by-products of railroad routes or highways.

Considered historically, this change in settlement formation is big enough to deserve the name "transformation." Almost always in the past (the exceptions may lie in civilizations like ancient Rome's), human settlement was a far slower and ecologically more intimate process. Everywhere, in the folk dimension, settlement was saturated by familiarity with nature and embedded in culture; but the new American settlements were increasingly the products (or by-products) of power policies enacted by civilized governance with massively accrued industrial capital dominating nature with a new magnitude of arrogant "resource" presumption and scorning culture as the effeminate preoccupation of insecure personalities. Indeed, rural culture is still considered an anachronism, a vestige or mere footprint of the past. Its crooked footpaths are to be superseded by the efficiency of sidewalks and the solidity of cement.

This conviction of the irrelevancy of rural culture is so normative in industrial ideology, and the condition of rural culture so debilitated, that partisans of rural life have an extremely difficult task in simply bringing the issue of rural decay into public awareness. As cities enlarged, the countryside shrank in importance. It's hard to think about things for which we have no thought. But rural culture must not be utterly dismembered; we need clear and decisive analysis as to its meaning, function, and validity.

In the following pages, I shall try to give expression to a bundle of ideas, values, insights, and beliefs related to the dangers inherent in the deterioration of rural culture. Though sometimes rude of manner and rough in expression, these essays represent an effort by a concerned citizen and a worried partisan of rural culture to shepherd an unruly flock of convictions toward a more complete understanding. Some issues can be readily named: the current industrial ideology of overproduction with its advertising "industry" geared toward a constantly increasing consumption of all marketable commodities; the tremendous shrinkage of agricultural society and the decimation of an integrated and relatively independent rural culture with its own folk traditions, crafts, tools, dance, music, and art; the industrialization of agricultural methods; the expansion of suburban lifestyles; endless urban sprawl with its highways, automobiles, gas stations, motels, and fast-food chains; an intensified preoccupation with the young, especially in regard to educational curricula bent on maximizing industrial science; the lowest possible denominator of entertainment via television (the democratic ideal of a cultural "melting pot" reduced to technological absurdity); human relationships trivialized as personnel within bureaucratic organizations in which one's time, energy, and even one's identity belong to impersonal bureaucratic forces; the absence of deep and sustaining friendships and communal vitality; a military-corporate partnership so entangled with the overall economy, as well as with the supposedly free universities, that its maintenance and expansion are seen as key and essential factors in continued industrial development and economic growth; a war machine of infantile masculine fantasy geared to Mutually Assured Destruction (MAD) and, more recently, to the illusion of a "protective shield" (Star Wars); and a psychotherapy that, finally catching up to the political theories of John Locke, preaches contract as the basis of interpersonal conduct.

Given the depth and intensity of these conditions and crises, it would be foolhardy to propose precise blueprints for how things *ought* to be. But if one takes the time and trouble to look for them, it's possible to find guidelines for a more peaceful and convivial world. Some of these guidelines can be found within, in the form of intuitions and esthetic judgments; some can be found without, in spiritual teachings and literary traditions. Any society that lived within real ethical boundaries and followed actual environmental limits would be far healthier, more beautiful, and more genuinely stable than our society presently is or perhaps ever was.

Blueprints and organizational charts are out of place in our envisioning a liveable future. The "good life" does not emerge from practiced technique but, rather, from ethical committedness and ecological sensitivity. The late philosopher Baker Brownell, friend and associate of Frank Lloyd Wright and author of *The Human Community*, focused repeatedly on the need for the restoration of rural community. This is what Brownell said:

> [C]ommunity cannot be manufactured. It cannot be built like a house. Though intelligence is needed to maintain it, the community itself comes, like life, without machinery or artifice. For the community is not formulated for power, profit, wages or production. It is the integrity of living.
>
> This integrity may not be deliberately planned. Planning can only improve the conditions under which communities may exist. It may be necessary to their survival. But the community is not this planning. Life under wholesome conditions has a way of assembling itself in a coherent pattern. It has what may be called organic intelligence . . . [1]

As we attempt to discover the nature of the wholesome conditions to which Brownell alludes, hoping to arrive at some specific answers, we can begin by realizing that small-scale village life, based on subsistence agriculture and handicrafts, has been the cultural root and stabilizing force within the history of civilization. This understanding has been elaborated upon by such thinkers as Martin Buber in his *Paths in Utopia*, E. F. Schumacher in *Small is Beautiful*, and Lewis Mumford in *The City in History*. In these books, one finds similar observations on the destructive power of excessive centralized control. Their primary answer to the problems created by excessive centralization is social, political, economic, and cultural decentralization—and decentralization is an empty term unless it implies the restoration of rural coherence.

All this merely sums up. Believing these principles to be true does not, in and of itself, create a liveable society. But a commitment to these principles can help bring about clarity of purpose. It helps one sort through the rubbish can of ideas. Principles are not, as Brownell recognized, the mental machinery out of which community can be fabricated. Community cannot be contrived or invented, as he rightfully insists. Community comes "like life, without machinery or artifice."

II

There are two radically different perspectives in regard to our immediate future, and the choice one makes between them is heavily determined by one's belief or disbelief in the promises of the industrial economy, science, and technology. The first alternative is that civilization, short of major alterations, is in for unprecedented turmoil; the second is that technology is firmly in command and that science will find a way out of any and all predicaments, even including the depletion of crude oil and the warming of Earth.

It's increasingly evident, however, that if science finds ways to maintain some semblance of industrial growth (by bringing yet more nuclear power plants on line, for instance), that will in no way address or resolve the fundamental problems of our age. For our root difficulties—the lack of coherent rural culture, uncontrolled urban sprawl, rapacious consumption, limitless industrialization, excessive regimentation in school and factory, indiscriminate use of chemicals, military overkill—are themselves the products of an inappropriate reliance on economic novelty, science, and technology as substitutes for cultural wisdom, spiritual humility, and social democracy. Another increase in energy production and commodity consumption will only aggravate the already explosive problems. Global warming is now a household term and peak oil is becoming one. It may be technically possible for "cities" to be placed in orbit in the twenty-first century, or for colonies to be established on the moon or elsewhere. But these futuristic scenarios, too, are no solution; they only add to and distract from the main difficulties. When Ronald Reagan said "Where we're going we won't need roads," it was hard to know whether he was talking about space travel or dystopian collapse. Until we have demonstrated both a willingness and an ability to live peacefully and ecologically on this planet, we have no right to spread our violence, confusion, and disorder throughout the solar system.

Civilization, lured by a fantasy of immortality and crazed by the mirage of progress, slips further and further from its evolutionary and historical roots. The roots of civilization—broad-scale, decentralized rural culture and folk community—have been pulled out of the ground by modern industrial civilization and beaten into commodity-intensive obedience. Our great contradiction is precisely that social stability and ecological coherence must be based on just such rootedness as civilization, especially industrial civilization, has sought to exterminate. In working for cultural renewal and the reconstruction of the countryside, we must be aware that *we are struggling directly against* the primary thrust of industrial civilization. Science and statistical economics may find ways to keep international capitalism grinding along; but it will grind on only at the price of environmental degradation and social desiccation. The political wreckage of society and the economic devastation of Earth must eventually stop. Earth and human nature cannot sustain such prolonged degradation, such intense disgrace. Most seriously, we are faced with the deployment of real weapons whose use could kill all of us and leave Earth a charred, radioactive, and burnt-out planet: Sigmund Freud's "death instinct" proven beyond the shadow of a doubt. The more dependent we become on the technocratic system, the more we are captivated by its promises and allurements, the more it replaces common culture, the more difficult it is to change in a truly positive manner and the more likely the system's breakdown will result in unparalleled destruction, carnage, and chaos. Despite a growing uneasiness among many of us regarding pollution, environmental damage, urban sprawl, an underpopulated countryside, and military intervention on behalf of an economic ideology that falsely parades itself as *conserve*-ative, the official commitment to industrial growth, and to the required energy consumption by which to promote that growth, remains unchanged.

One thread that links our present industrial affluence to our growing international hostility is energy: a boundless consuming gluttony of oil and electricity that hyperactivates our economy, our foreign policy, and our personal lives. The whole industrial world, with the United States leading the way, is high on energy. The remark by the late E. F. Schumacher, in his *Small is Beautiful*, that our "present consumer society is like a drug addict" is both succinct and substantially correct.[2] The possibility that our "drug" might be cut off by economic collapse or radically reduced by ecological exhaustion or military catastrophe raises our alarm and intensifies our hostility—although it might be said that the real drug, for most people, is not energy addiction per se but consumer habituation. Our task, and it is no small one, is to squarely face our destructive comfort addictions and take the necessary (if also painful) steps to get off the hook. Habit creates its own justification; and as our present system enervates common culture, so we have been taught to despise the past and to fear serious and productive cooperation. Thus we stand ethically incapacitated and morally numb in the midst of pressing need. Our envisioning of a democratic and ecologically sound future is, therefore, both an urgent and a concrete task, although we must not let such envisionings become mere idyllic daydreams, pleasant but powerless.

It is customary, when writing about urgent social, economic, and political dilemmas, to let in an unexpected ray of light, to be quietly optimistic, to suggest that our "leaders" will wake up in time, that a little more education will provide the solution. Such optimism relieves one of personal responsibility; it helps us sleep better at night; it suggests that those who know (or who should know) will act, in the end, wisely and well. This is an optimism in which we can no longer indulge. The renewal of cultural values is a grassroots struggle or it is nothing at all. Civilization tends to be identified with its leaders and heroes; but those who promise to advance its progress must first pass, in our electoral system, through the ideological filters of the corporate elite, thus becoming dependent on their financial largesse and bound to their economic interests. To wait for officially sanctioned leadership is to wait for the tomorrow that never comes. We need a new political force from the grassroots, a coalition that will lay the compelling issues on the line, compose a clear and coherent platform, and offer its own candidates for public office. The most promising constituencies are those that go by the names "Green" and "Rainbow." But such a political force must be built on the strength of local communities and networks, and on the growing participation of those previously uninvolved people for whom politics seems merely the hyperactive pastime of the overly educated unemployed.[1]

Neither "major" political party in the United States offers a structural alternative to industrial capitalism. As the "opposing" parties staked out their positions prior to the 1984 elections, for instance, both Republicans and Democrats attempted to outbid each other, as always, in the promotion of high-technology progress. If Republicans stand for Research and Democrats for Development, we need a political gathering that represents something more significant than capitalist R&D. Walter Mondale went so

---

1. It must also be said that when peace candidates (like Congressman Dennis Kucinich) make themselves available in presidential races, and when such persons are derisively labeled "unrealistic" and "unelectable," and so cannot attract sufficient voters or financial support to sustain their candidacies, then we must also recognize that mainstream politics is a rough reflection of main street self-satisfaction. There may not be a perfect correlation here, but neither is there a total disconnect. Public shallowness and elite concept mongering are reciprocal enablers for the maintenance of the status quo.

far, as he announced his candidacy for the office of President, to say that "Science must teach us the future." Such a pronouncement raises intellectual pandering to the level of political program.

III

Let us listen to a wiser voice speaking on the subject of science. The author is the late Swiss psychiatrist Carl G. Jung. The following passage comes from a book, recorded and edited by Aniela Jaffe, entitled *Memories, Dreams, Reflections*:

> Our souls as well as our bodies are composed of individual elements which were already present in the ranks of our ancestors. The 'newness' in the individual psyche is an endlessly varied recombination of age-old components. Body and soul therefore have an intensely historical character and find no proper place in what is new, in things that have just come into being. That is to say, our ancestral components are only partly at home in such things. We are very far from having finished completely with the Middle Ages, classical antiquity, and primitivity, as our modern psyches pretend. Nevertheless, we have plunged down a cataract of progress which sweeps us on into the future with ever wilder violence the farther it takes us from our roots. Once the past has been breached, it is usually annihilated, and there is no stopping the forward motion. But it is precisely the loss of connection with the past, our uprootedness, which has given rise to the 'discontents' of civilization and to such a flurry and haste that we live more in the future and its chimerical promises of a golden age than in the present, with which our whole evolutionary background has not yet caught up. We rush impetuously into novelty, driven by a mounting sense of insufficiency, dissatisfaction, and restlessness. We no longer live on what we have, but on promises, no longer in the light of the present day, but in the darkness of the future, which, we expect, will at last bring the proper sunrise. We refuse to recognize that everything better is purchased at the price of something worse; that, for example, the hope of greater freedom is canceled out by increased enslavement to the state, not to speak of the terrible perils to which the most brilliant discoveries of science expose us. The less we understand of what our ancestors sought, the less we understand ourselves, and thus we help with all our might to rob the individual of his roots and guiding instincts, so that he becomes a particle in the mass, ruled only by what Nietzsche called the spirit of gravity.

Reforms by advances, that is, by new methods or gadgets, are of course impressive at first, but in the long run they are dubious and in any case dearly paid for. They by no means increase the contentment or happiness of people on the whole. Mostly, they are deceptive sweetenings of existence, like speedier communications which unpleasantly accelerate the tempo of life and leave us with less time than ever before. *Omnis festinatio ex parte diaboli est*—all haste is of the devil, as the old masters used to say.

Reforms by retrogressions, on the other hand, are as a rule less expensive and in addition more lasting, for they return to the simpler, tried and tested ways of the past and make the sparsest use of newspapers, radio, television, and all supposedly timesaving innovations.[3]

Jung has here proposed an extraordinarily wise social and political principle: that we learn to balance our scientific and technological advances with appropriate cultural retrogressions. He is saying that to counter the industrial assault on the future requires a serious and stubborn determination to live more "slowly" in the present, with greater reliance on both the wisdom and methods of the past. It would, however, be a grave mistake to equate "reforms by retrogressions" with a wholesale return to so-called primitive conditions. Reform by retrogression is neither a simple nor a simple-minded concept. To live directly on the energy of the sun, for instance, requires a great complexity of ecological and cultural sophistication. Jung clearly intended the idea of retrogression as a means by which we could renew a badly needed contact with the past, with community, and with nature. He saw that modern life—the "cataract of progress"—has removed us from the natural world in ways that have given rise to massive "discontents." Explicit in this view is the insight that civilization's discontents are only exacerbated by the continual acceleration of progress.

A partial list of industrial civilization's discontents can be found in "The Wages of Growth," an essay by the British economist Ezra J. Mishan, in a book entitled *The No-Growth Society*, edited by Mancur Olson and Hans H. Landberg. According to Mishan:

Human contacts decline with the spread of labor-saving and time-saving machines. They decline with the growth of supermarkets, cafeterias, vending machines, private cars and airplanes, with the spread of transistor radios and television sets, with computerization in offices and patient-monitoring machines in hospitals, with closed circuit television instruction and teaching machines. Sanctioned by a restricted interpretation of economic efficiency, the main thrust of consumer innovation since the turn of the century appears to have been directed toward producing for us a push-button world in which our trendy whims are to be instantly gratified while our psychic needs are increasingly thwarted. The vision of such a world—a universe humming with recorded instructions and electronic devices that will herd the multitude along moving belts and through sliding doors, that will feed and tend and lull each one of us without so much as a human twinge—may inspire the technocrat and elate the growthman. But the unavoidable consequence is that the direct flow of sympathy and communication between people becomes ever thinner. And to that extent the quality of their lives becomes ever poorer.[4]

In the passage just quoted, Mishan concentrates on the internal quality of what we might call the public community, not on the strength of contact between people and nature. Yet they—community and the environment—are inextricably linked: anything that corrupts human community, including those future shock "improvements" to which Mishan refers, will have an injurious impact on our relationship with nature, even if temporarily disguised or rationalized as unintended "externalities." The more removed we become from nature, the more inclined we are to vacillate between sentimentality and indifference in regard to Earth's ecology. Only a society that maintains a strong, functional, and respectful relationship with the natural world is capable of adequately caring for Earth in the long run, of keeping its economy within ecological limits.

It's true that to be alive is to influence and meddle with the environment. To be human is to intrude upon the web of life in radically different ways than plants or other animals. But different cultures have left vastly dissimilar records of their relationship to Earth. Gathering and hunting cultures, for instance, have had a far gentler impact on native ecology than those cultures that took to farming. And in those cultures where farming led to the rise of civilization—a development we will turn to shortly—the civilized system, in turn, imposed demands upon the primary producers that led to increasing insensitivity and exploitation. The ancient relationship between culture and nature was strained and broken by the emergence of a civilized system with vast power to command and compel. Those people who acted directly on the environment as farmers, loggers, miners, and so forth, were no longer engaged in modest subsistence activities in behalf of a self-provisioning community; rather, they were extracting an immodest *surplus* for urban (or, at least, aristocratic) consumption. This surplus could only be obtained from the environment itself, by disrupting nature in an unprecedented way. Sustained exploitation, in turn, served to distance working people from their traditional respect for Earth. Social change means cultural change; to become civilized requires ecological insensitivity.

Yet it has been our peculiar freedom and genius as human beings to control fire, construct shelters, fabricate clothing, learn the healing arts, invent tools, plant fields, domesticate animals, build roads, and so on. The human record is rich with discovery and invention. Innovations have multiplied, generation after generation, especially in the last two hundred years. But the rate of change is now so great, and the predatory economic motive so ascendant, that virtually all traditional caution has been superseded by the reckless pursuit of progress. There may be little point in objecting to such inventions as electrical power and the automobile, in and of themselves. It is rather the lack of careful discrimination, the boundless gluttony, and the brazen abandonment of crafts and patterns of cooperation that have caused us to lose contact with our essential cultural heritage and with Earth itself. This country did not have time to assimilate the railroads before the automobile system burst on the scene with endless miles of asphalt and concrete, making the railroad and its effects seem mild, quaint, and charming by comparison. If we ask ourselves, in quiet seriousness, whether we are culturally richer or poorer for having discarded railroads, sailing ships, steam boats, even horses and buggies in exchange for automobiles and intercontinental jets, the answer—if we are honest both to our intuitions and to the renewal power of Earth—is that we are by no means better off. To ask such questions, however, even granting the usefulness of cars and jets, is to provoke only an amused contempt from those who stand in command of the machines of progress.

Progress addiction notwithstanding, it remains true that the wholesale elimination of the "tried and tested ways of the past" leaves us exactly with what Carl Jung describes: a vastly accelerated life pace that works directly against natural attentiveness, ecological sensitivity, and peace of mind. That is why it is important to grow vegetables and raise chickens, to have a hand in building one's own home, and to watch the wood burn in one's own hearth. How fully one partakes of such retrogressions is invariably a matter of personal preference and ability—except for those people, of whom there are a great many, whose very poverty excludes them from participation in active self-provisioning. (This is a great and tragic irony within industrial society, for in the preindustrial past the rural poor of necessity lived a life of cooperative subsistence in the context of a folk commons whose existence was thought to be guaranteed in perpetuity. In the present dispensation, however, when virtually all existence is both formally legal and structurally private, the poor, now displaced from both nature and the commons, have no alternative but governmental handouts or, increasingly, minimum wage jobs while living in the bleakest possible circumstances. These poor people constitute a subclass in society and are seen as a sluggish and misbehaving burden—barely human "bottom feeders.")

It is nevertheless clear, precisely with a powerless subclass in the midst of so much affluence, that the continued rejection of reforms by retrogression gives us much that is intolerably sterile in modern life and that effects the containment of an earthier connectedness. In his book *One-Dimensional Man*, Herbert Marcuse says that "To a great extent, it is the sheer *quantity* of goods, services, work, and recreation in the over-developed countries which effectuates this containment. Consequently, qualitative change seems to presuppose a *quantitative* change in the advanced standard of living, namely, *reduction of overdevelopment*."[5] Or, as Marcuse puts it in his earlier work, *Eros and Civilization,* " . . . the regression to a lower standard of living . . . does not militate against progress in freedom . . . The argument that makes liberation conditional upon an ever higher standard of living all too easily serves to justify the perpetuation of repression."[6] Our present standard of living not only serves to justify human repression, it also serves to justify the degradation of Earth. It is precisely on the suppressed energy of this repression that advertising plays: with this brand of cigarettes we can be uninhibited; with that beer we can live "in God's country." Our fabulously high industrial standard of living, in the context of a world that contains unbearable poverty and deepening ecological nightmares, is based on both human and environmental exploitation. We have democracy for capitalists and socialism for the rich; we have slums for the poor and toxic waste dumps for Earth.

Rightly understood, a higher standard of living *is* needed—not on the completely misleading basis of sheer technological quantity as registered by the Gross National Product (which, as some wag has remarked, always manages to live up to its name), but rather in the quality of communal life and stability of environmental coherence. As Aldo Leopold said in *A Sand County Almanac,* " . . . our bigger-and-better society is now like a hypochondriac, so obsessed with its own economic health as to have lost the capacity to remain healthy."[7]

Our health, both culturally and socially, will be restored only when we stop abusing the environment, provide real living opportunities for the poor, and adopt clear policies of rural reconstruction. But until those changes occur, our overall condition can only be diagnosed as malignant proceeding toward terminal. Such is the prognosis of unrestrained progress. Such is the fate of our "mastery" over nature.

ENDNOTES

1. Brownell, *Human*, 98–99.
2. Schumacher, *Small*, 153.
3 Jung, *Memories*, 235–37.
4. Mishan, *No-Growth*, 85–86.
5. Marcuse, *One-Dimensional*, 242.
6. Marcuse, *Eros*, 153.
7. Leopold, *Sand*, xix.

2

Nature's Unruly Mob

T HE STATISTICAL DATA ON farms in the United States indicate a continued growth in the size of farms and an ongoing decline in the
number of farms. Between 1940 and 1960, nearly twenty million people
left their farms. During the same period, the number of farms dropped
from around 6.5 million to less than 2.5 million. Despite a remarkable
and unprecedented migration turnaround during the 1970s—about three
million more people moved *into* rural areas than *from* them—net farm
loss continued to accelerate. Actual farm population dropped by one-
third in the 1970s. In 1970 there were, according to census figures, 8.3
million people on farms; by 1980, the number of people on farms in the
United States had dropped to 5.6 million. In the mid-1980s, farm population constituted approximately 2.2 percent of the national total—an unprecedented low. These data, with their austere numerical features, clearly
indicate the continued decline of rural culture.

At least a dozen states, again in the mid-1980s, reported less than one
percent farm residents; these included New Hampshire, Massachusetts,
Rhode Island, Connecticut, New York, New Jersey, Florida, Arizona,
Nevada, California, Alaska, and Hawaii. In the 1970s, in the northeast,
Vermont could still boast a 3.5 percent farm population. In the south,
Kentucky had 6.7 percent farm residents and Arkansas 4.7. In the west,
Montana and Idaho had 7.4 and 7.3, respectively. And, in the upper
Midwest, both North and South Dakota had over 16 percent, Iowa had
13.4 , Nebraska 11.7, Minnesota 7.7, and Wisconsin 6.0.

The ongoing drain of farmers from the social fabric, and the corresponding depletion of rural culture, is the direct result of economic
pressures and political policies associated with the growth of industry
and commerce. This economic and political process began to accelerate
in Europe several centuries ago, became institutionalized in a new way

through the Industrial Revolution, and was officially "justified" by the emergence of economics as an industrial "science." The sharp decline in rural population has not only been called inevitable by industrial economists, it has also been hailed as a civilized advance. Small-scale, labor-intensive farming is the very model of "backwardness" that industrial planners now seek to uproot in all so-called underdeveloped countries. An "underdeveloped" country is typically an *agricultural* country with a vast rural population composed heavily of small-scale and subsistence-oriented farmers. To be "brought into the modern world," such a country is to force itself, ruthlessly if necessary, out of its old rural culture and into a new urban-industrial way of life. This is what is meant, in essence, by "development." What we are not told by its promoters is that "development" is the latest, and perhaps the last, step in the civilized attempt to surpass peasant, folk, and rural culture. When indigenous cultures are all corrupted and the last peasant is dead, we will all celebrate our universal civility in ubiquitous boredom and poisonous sterility.

What development typically brings with it is an enforced breakdown of indigenous culture with the subsequent atomization, identity loss, mass loneliness, heightened malnutrition, increased unemployment, mounting crime, and widening wealth chasm so common in the history of most industrialized countries. As more and more people are forced off the land, new techniques for industrial "farming" can be imposed with little regard for the "superstitious" protests of native inhabitants harkening to lore, tradition, spiritual values, and a lifelong intimacy with soil. Nature becomes increasingly vulnerable to civilized manipulation. To "rationalize" agriculture is to apply civilized techniques directly to nature's body. In earlier times, subsistence-oriented peasants to some degree buffered nature from excessive abuse. A purely commercial agribusiness pursues maximum yields, irrespective of ecological or cultural consequences. In this country, we have virtually no cultural memory of the folk commons; and in some parts of the country, like my own in northern Wisconsin, small-farm culture lasted only a generation or two before being wiped out by "economies of scale."

II

There are additional considerations and implications in this way of looking at history. First, there is a profound and inevitably disastrous imbalance in urban versus rural populations—and by rural I mean primarily, but not solely, that portion of the population that is agricultural. This population imbalance worsens in proportion to the enlargement of industrial agribusiness; and, in some instances, the depth of rural crisis can be glossed over by misleading statistics that fail to distinguish farm from nonfarm households.

The concentration of industry in metropolitan areas, with the attendant massing of human populations, has caused serious damage to agricultural life—the culture of the countryside—on a worldwide scale. Efforts to call agriculture an industry, to rename agriculture "agribusiness," have already succeeded. As E. F. Schumacher has stated in his *Small is Beautiful*:

> The farmer is considered simply as a producer who must cut his costs and raise his efficiency by every possible device, even if he thereby destroys—for man-as-consumer—the health of the soil and the beauty of the landscape, and even if the end effect is the depopulation of the land and the overcrowding of cities . . . In our time, the main danger to the soil, and therewith not only to agriculture but to civilization as a whole, stems from the townsman's determination to apply to agriculture the principles of industry.[1]

As Schumacher goes on to say:

> [T]he fundamental 'principle' of agriculture is that it deals with life, that is to say, with living substances . . . The fundamental 'principle' of modern industry, on the other hand, is that it deals with man-devised processes which work reliably only when applied to man-devised, non-living materials . . . *The ideal of industry is to eliminate the living factor* . . . (Emphasis added.)[2]

These observations on the *fundamental* difference between agriculture and industry lead E. F. Schumacher to a truly powerful insight, an insight that requires nothing less than an entirely fresh reappraisal of civilization:

> [T]he fundamental 'principles' of agriculture and of industry, far from being compatible with each other, are in opposition. Real life consists of the tensions produced by the incompatibility of opposites, each of which is needed, and just as life would be meaningless without death, so agriculture would be meaningless without industry. It remains true, however, that agriculture is primary, whereas industry is secondary, which means that human life can continue without industry, whereas it cannot continue without ag-

riculture. Human life at the level of civilization, however, demands the *balance* of the two principles, and this balance is ineluctably destroyed when people fail to appreciate the *essential* difference between agriculture and industry—a difference as great as that between life and death—and attempt to treat agriculture as just another industry . . . What is involved is the whole relationship between man and nature, the whole life-style of a society, the health, happiness and harmony of man, as well as the beauty of his habitat.[3]

If we merge Schumacher's principles of agriculture and industry with Carl Jung's principles of reform-by-retrogression and reform-by-advance, we have some truly powerful concepts with which to forge a new cultural agenda. First, we can identify the living, growing, biological nature of true agriculture (onto which an industrial, bioengineered economy has been pressed), and, second, we can recognize the profound psychological and spiritual necessity for a culture rich in the "tried and tested ways" of the past. The issue then becomes, as Schumacher is quick to point out, a matter of balance. By promoting only industrial progress, modern society enlarges the sphere of industrial consolidation and constricts the sphere of agricultural coherence. (One must be careful, however, not to place agriculture always and only with retrogression, for many aspects of what "parity" advocate Charles Walters has called "eco-agriculture" would properly fall under the principle of advance. On the other hand, there are advances to be made in solar and wind technologies, as in lightweight mass transit; and the building trades, falling roughly under the heading of industry, are in sore need of craft revival.)

If Schumacher is correct when he insists the "whole life-style" of our society is at stake, specifically in our failure to appreciate the *essential* difference between agriculture and industry, then we would do well to ponder the implications of our present system and reconsider the destination of progress: the ideal of industry, says Schumacher, is to *eliminate* the "living factor." The continued maintenance of nuclear weapons clearly confirms this assertion, and their use would ratify it unconditionally. As Schumacher goes on to say:

The social structure of agriculture, which has been produced by— and is generally held to obtain its justification from—large-scale mechanization and heavy chemicalization, makes it impossible to keep man in real touch with living nature; in fact, it supports all the most dangerous modern tendencies of violence, alienation, and environmental destruction. Health, beauty, and permanence are hardly even respectable subjects for discussion, and this is yet another example of the disregard of human values—and this means a disregard of man—which inevitably results from the idolatry of economism.

If 'beauty is the splendor of truth,' agriculture cannot fulfill its . . . task, which is to humanize and ennoble man's wider habitat, unless it clings faithfully and assiduously to the truths revealed by nature's living processes. One of them is the law of return; another is diversification—as against any kind of monoculture; another is decentralization, so that some use can be found for even quite inferior resources which it would never be rational to transport over long distances. Here again, both the trend of things and the advice of the experts is in the exactly opposite direction—towards the industrialization and depersonalization of agriculture, towards concentration, specialization, and any kind of material waste that promises to save labor. As a result, the wider human habitat, far from being humanized and ennobled by man's agricultural activities, becomes standardized to dreariness and even degraded to ugliness.[4]

All these observations lead Schumacher to this profoundly useful proposal:

> Instead of searching for means to accelerate the drift out of agriculture, *we should be searching for policies to reconstruct rural culture*, to open the land for the gainful occupation of larger numbers of people, whether it be on a full-time or part-time basis, and to orientate all our actions on the land towards the threefold ideal of health, beauty, and permanence. (Emphasis added.)[5]

We should be searching for policies to reconstruct rural culture! This is exactly the key issue for the revival of Western society as we enter the twenty-first century.

Quite obviously, the "new rural culture" Schumacher envisions and advocates will arise only through the voluntary participation and political action of the citizenry as a whole. Thoughtful governmental policy could enable this reconstruction to proceed with alacrity. The present rural population cannot effect such a transition on their own; it is simply beyond their power to do so. Those remaining in agriculture are typically well past middle age, and a great many "farmers" are agribusinessmen by economic attrition. To go beyond the sterility of the existing system requires that the principles and concepts of "Green" politics must undergo wide acceptance and support: it will take a heady mix of city people intellectually familiar with advances in appropriate technology and organic methods and those remaining country people still in touch with older agricultural practices to bring about this depth of change. In addition, and most dismaying, the program of civilized progress is so technologically self-assured and historically arrogant that the renewal of the countryside at this point probably depends, paradoxically, on some degree of systemic breakdown—a prolonged financial crisis, such as the Great Depression, or a fundamental energy shortage.[1]

It is in this area where the large environmental organizations could exert a great deal of influence, if they could only see that environmental protection is doomed to be merely a hopeless holding action unless, as a whole society, we squarely face and come to grips with technological growth and begin to reconstruct rural culture. Environmental preservation is finally impossible without a strong and stable rural culture. It is disheartening to see how few environmentalists seem to understand this dilemma.

Perhaps an anecdote is in order here. In the campaigning prior to the 1980 presidential election, environmentalist Barry Commoner, running on the Citizens' Party ticket, spoke in Indiana. After his speech, I had the opportunity to ask Commoner a few questions. First I asked him how large an increase in farm population would be required if a biological agriculture were to be restored. Commoner said he had never really thought about it. Well, I said, we presently have less than four percent of the total population involved directly in farming. He thought a moment and said: about a fifty percent increase. So, I replied, you would call for an increase from four to six percent? Yes, he said, that seems about right. I then asked him what he thought about E. F. Schumacher's proposal for reconstructing rural culture and, furthermore, whether such a reconstruction was necessary to move us beyond the dangers of rampant industrialization. This time Commoner did not hesitate. Schumacher, he said, was always full of clichés.

Until environmentalists, including such fine and forceful leaders as Barry Commoner, realize that the environment cannot be substantially

---

1. In the past few years the terminology for such woes has become increasingly commonplace: terrorism, global warming, climate change, peak oil.

protected unless industry itself is brought under control, and that, accordingly, industry will never be brought under control until the renewal of rural life is encouraged and politically promoted, we will see little change for the better in the condition of Earth. It is not enough to speak in general terms about the "environment." We must speak *for* the land and live *on* and *with* the land. A stable society needs an embedded rural culture.

III

One of the best known advocates for small-scale organic agriculture in the United States was Robert Rodale. Rodale was the author of several books, chairman of the board of Rodale Press, and editor of *Organic Gardening*, *Prevention*, and other periodicals. In an essay entitled "Breaking New Ground: The Search for a Sustainable Agriculture" (that first appeared in the February 1983 issue of *The Futurist*), Rodale had this to say:

> One of American agriculture's most serious short-term economic problems is surplus production. But those unneeded piles of corn and other crops are not merely temporary signs that the weather has been too good; they are a sign of a basic fault in the fundamental concept of our farming system. Crop and commodity surpluses, which many point to as a force driving down prices, are not a *primary* problem of our agriculture . . . [F]rom my perspective they are the *result* of a problem, and not the problem itself.
>
> The fault that surpluses signal is this: Farmers are using methods that do not allow flexibility in the amount they produce. Conventional American agriculture works only when the throttle governing energy and input flows is pulled all the way out. Farmers lack the option of switching easily—either permanently or temporarily—to an alternative system that works well with very low levels of inputs, yet would still provide reasonable levels of production permanently (without erosion).[6]

Although Rodale does not really address the issue of *why* farmers "lack the option" to switch to an alternative system—Rodale is practically mute on the subject of politics—he does say "The attitude that more food is always better has led us toward a strategy of dominance in our approach to nature." He also says that, in the last two centuries, the "capacity of farmers to dominate natural forces grew dramatically as the availability of inputs increased."[7] This new capacity to dominate nature was due to a number of factors:

An important turning point was the work and thinking of Jethro Tull, an English farmer and inventor. In 1733, Tull published his well-known book *The New Horse-Houghing Husbandry*, and within 50 years the ability of farmers to dominate nature had increased dramatically.

Tull was so influential because he created a whole new farming system. Before Tull, farmers had neither the ideas nor the technology to plant their seeds in a way that would give crops the best chance to compete against weeds. They plowed, but they did not have machines to break up clods thoroughly. Tull invented and built those machines.

Tull also invented the first seed drill and showed farmers how to plant their crops in rows instead of just broadcasting seeds. With crops planted in rows, it became possible to cultivate to control weeds. Tull invented cultivators, too. In short, the whole concept of thorough tillage, row cropping, and keeping the soil surface as bare as possible emerged from the brain of Jethro Tull.[8]

Tull's ideas and innovations, according to Rodale, not only "enabled a great increase in food production," they also advanced the "ability of farmers to dominate nature."[9] Rodale goes on to say that Tull

... showed farmers how to place crop plants in their fields in such a way that the line between nature's area and that of the farmer's was more clearly drawn. Nature's area was the space between the rows—and that was swept as clean of life as possible. The farmer's territory was, as always, the crop plant itself and the land it occupied. Because it was now more separate from nature's unruly mob of weeds, the farmer's dominion could thrive.

Others, following Tull, took many more steps toward an era of agriculture in which the domination of people over nature became almost complete ... [O]ne of the most important was Justus von Liebig, who set the stage for the chemical domination of the soil through the extensive use of fertilizers.[10]

Tull's method for controlling nature's unruly mob of weeds was exported to the United States; but, according to Rodale, the more violent American rains aggravated an old agricultural problem:

Erosion became a major force, depleting the character and quality of the soil and reshaping the face of the earth. Today, nearly six billion tons of topsoil wash and blow away each year. At that rate of soil loss, there is no way that farming in the United States can be sustained on a permanent basis.[11]

Tull's techniques for planting and cultivating also set the stage, says Rodale, for agriculture's "dependence on the regular flow of fertilizers, pesticides, fuel, and non-replaceable minerals."[12] These remarkable gains in crop yields have been heralded by agribusiness concerns and industrial economists as incontrovertible proof of the superiority and efficiency of the new agro-industrial techniques. But that assertion, too, is highly questionable:

> Before the time of Jethro Tull, agriculture almost everywhere was an energy-collecting system, drawing on the free energy of the sun for almost all the force needed for production. Tull, by showing people how to dominate nature more effectively, made the river of agri-energy flow uphill, so to speak. *Before, the farmer got back perhaps 10 calories in food for every calorie invested into the agricultural process. Now, for every 30 or so calories we pour into agriculture in the form of fossil energy, we get back only one.* (Emphasis added.)
>
> The domination approach to agriculture is leading farmers toward self-destruction, both economically and biologically. The desire to dominate nature causes farmers to use excessively large amounts of nonrenewable resources, which are continually increasing in price. And the almost exclusive orientation of our research policy toward increasing the ability to dominate has led farmers away from the production flexibility they need to be able to pull back economically during periods of surplus production. Finally, the domination policy has placed soil protection at a low priority. In that sense, the domination concept is a sham. Farmers think they are dominating nature, but they are doing that only while they have soil.[13]

Organic farming is one of the "alternatives to the road toward dominance."[14] Rodale says that by

> ...avoiding the use of synthetic fertilizers and pesticides, a farmer can give up the basic tools of dominance and exchange them for more natural fertilizers and totally biological approaches to insect control...
>
> Eventually, the role of organic farmers should be to venture entirely beyond the present system toward a much broader and potentially more valuable concept I call regenerative agriculture ... The regenerative farmer would move much further from the dominance idea, and instead use a production system depending to a very high degree on the free goods that nature provides. Farming would then change from a battle against nature into the art of encouraging nature to release the most benefits for human use with the least possible effort.[15]

Rodale's prose is deceptively quiet and unruffled. Yet the seemingly casual string of ideas, observations, and points of historic interest adds up to a revolutionary whole. To move from the "dominance idea" in actual agricultural practice—and one of the key quantitative measurements of such a movement would be a net gain of caloric energy per unit input—would require not only massive changes in present agricultural methods, it would also imply a reversal of the whole agenda for maximizing yields while squeezing the life out of the countryside, both its soil and its people.

It can certainly be argued that Rodale never dealt directly with the reconstruction of rural culture. While that seems true, the renewal of rural life follows logically from the premises of regenerative agriculture. At minimum, a newly invigorated rural culture would be a "side effect" of such a major change in agriculture approach—although I think it both intellectually timid and politically irresponsible to treat cultural issues as mere "side effects" of technical implementations. There is simply no way to create a sophisticated biological agriculture that aims to work closely *with* nature that does not involve a significant reduction in farm size and a large increase in rural population.

The contraction of industrial agriculture implies the formation of a new standard for *all* industrial production. In other words, agriculture can become "regenerative" along the lines that Rodale proposes *only if our whole society undergoes a fundamental change of values.* Such new values are mostly old values taken seriously. We would not place profit-maximizing production, agricultural or industrial, in the position of high esteem it now commands; the GNP would cease to be the major measure of social importance and economic "health." A principle of regeneration would not promote individual or corporate gain at the expense of the environment, but would rather create a balance between social well-being and ecological stability. All this implies a reduction of scale, a paring down of inputs, a greater orientation toward local and regional production, an extensive decommercialization of food production and distribution (at minimum a growth of cooperatives involving producers and consumers without the monopolistic intervention of giant agribusiness firms and food processing corporations), a revival of self-provisioning, and a dramatic rise in rural population.

By using our industrial capacity and know-how wisely, for purposes that enhance our cultural richness with the least possible environmental intrusion, we could liberate millions of people from the boredom and drudgery of lifelong wage work and enable them to discover or rediscover the world of nature. We could significantly reduce the workweek. We could significantly increase the actual pleasure of creative labor. It is reasonable to assume that people would begin to take a fresh interest in life and the creative nurture of life—after passing, perhaps, through a phase of emotional lethargy, the logical result of lifting compulsive boredom. Many people would, if provided the option, participate in cooperative farm life or live in closer proximity to agriculture. The undistracted human spirit seeks depth and continuity. We need to restore contact, as Carl Jung admonished us, with those "components" in our deeper selves that "find no place in what is new." We will do that by living quietly and richly on the land.

ENDNOTES

1. Schumacher, *Small,* 106, 109.
2. Schumacher, *Small,* 110.
3. Schumacher, *Small,* 111–12.
4. Schumacher, *Small,* 114–15.
5. Schumacher, *Small,* 114.
6. Rodale, "Breaking," 15.
7. Rodale, "Breaking," 16.
8. Rodale, "Breaking," 16–17.
9. Rodale, "Breaking," 17.
10. Rodale, "Breaking," 17.
11. Rodale, "Breaking," 17.
12. Rodale, "Breaking," 18.
13. Rodale, "Breaking," 18.
14. Rodale, "Breaking," 18.
15. Rodale, "Breaking," 18.

3

Roots

Robert Rodale's concern for regenerative farming takes us to the leading edge of a new orientation toward regional, local, village, and household self-provisioning. This is no insignificant matter. It requires both personal commitment and collective political will to envision and articulate the necessary principles and to empower those convictions with personal and collective action. But short of such reorientation, we can expect increasing rootlessness as remaining small communities are disrupted, as culturally stable traditions are lost to or forgotten by newer generations indifferent to the past, as standardized urban mass culture continues to expand, as corporate hubris becomes more manic, and as military wrath reaches cosmic proportions.

World population will, according to the projections of demographers, go well over nine billion people before the twenty-first century is half gone. Weather patterns appear increasingly erratic, storms and droughts more severe. The Ogallala Aquifer in the southern Great Plains is being sucked dry by pumping for irrigation; fresh, clean water, free of pollutants, is everywhere becoming shorter in supply. Vast quantities of petroleum continue to be consumed with little attention given to conservation or to alternative and renewable sources of energy.[1] Pressure is strong from the military-industrial complex to maintain and even

---

1. In the spring of 2008, as this book was in process of being edited and revised for reprinting by Wipf and Stock, food prices were rising quickly, spurred by a combination of rapidly increasing oil costs and a federally supported effort to turn corn into ethanol. This alarming spike in prices should usher in a revived interest in and commitment to wind and solar energy, but we still have yet to face all the accrued cultural, economic, political, and religious resistances to reform by retrogression. No one is proposing an increased gas tax to fund an enlargement of public mass transit, nor does there appear to be any meaningful federal action to control rampant speculation in oil futures.

increase the number of nuclear power plants. Nuclear apparatus, both commercial and military, continues to exist even though most citizens, despite the official assurances of the Department of Energy, are unconvinced that there is any "permanent" method of or location for disposal of radioactive wastes—deadly in some isotopes for thousands of years. The production, deployment, and sale of military armaments grow apace. And nuclear "numbness" (information overload coupled with excessive affluence) protects us from taking seriously the insane destructiveness of even a "limited" nuclear war.

In agriculture we find that the industrial procedures adopted by American agribusiness permit a vast overproduction of domestic foodstuffs through monocropping mass production and chemical technology. Small-scale organic agriculture is officially derided as anachronistic or otherwise pigeonholed as a "niche market." The alarming rate of farm foreclosures indicates the tightening of the industrial-financial noose on agriculture; yet the underlying structural forces behind foreclosures are systematically passed over in favor of specious charges of "bad management" on the part of debt-ridden farmers. (In many instances the accusation is true—farmers *were* talked into taking on debt loads they could not carry—but such debt was entered into because of expansionist fantasies promoted through agribusiness ideology, with its lust for growth and giant toys.) Every decline in farm population erodes even the faint remains of rural culture.

Lacking functional ties to the natural world, society becomes ever more organizationally standardized while the young grope blindly for escape, knowing instinctively as human beings that they are being suffocated in organizational emptiness. The lack of cultural coherence is revealed in drug use by students even in elementary schools; teachers and students are sometimes required to wear identification cards; armed guards patrol the halls and stand sentry at the doors. The fear of crime keeps many city, and a great many rural, residents indoors at night; the television programs that serve as standard entertainment in advanced civilization are filled with oddly passionless spectacles of crime, sex, violence, murder, and police. According to Eugene H. Methvin, writing on television violence in the January 1983 Reader's Digest:

> Television ranks behind only sleep and work as a consumer of our
> time. In fact, according to the 1982 Nielsen Report on Television,

the average American family keeps its set on for 49 hours each week. The typical youngster graduating from high school will have spent almost twice as much time in front of the tube as he has in the classroom—the staggering equivalent of ten years of 40-hour weeks. He will have witnessed some 150,000 violent episodes, including an estimated 25,000 deaths.[1]

Jerry Mander, in his book *In the Absence of the Sacred*, says "It is hardly an exaggeration to say that the main activity of life for Americans, aside from work or sleep, has become watching television. Television has effectively replaced the diverse activities of previous generations, such as community events, cultural pursuits, and family life."[2] This is the creative content of advanced industrial civilization; this is what we do with our free time; this is the measure of our cultural paralysis, as we sit passively and stupefied before a flickering screen while the system saturates our consciousness with images of violence and sex. Such images are artfully shaped in order to maximize our vulnerability as consumers who will buy what we most certainly do not need, but by such buying are reaching, subconsciously, for some of the glamour, potency, and magic the artful images convey. This buying stimulates the economy and enlarges the GNP. Meanwhile, economists call for a continually expanding economy in order to hold inflation and unemployment within manageable bounds; they tinker with "controls" as if playing with a huge machine nobody knows quite how to run, a machine whose purpose drips with fantasies of omnipotence, power, and indestructibility.[II]

For all too many people, life in modern civilization is threaded with chronic boredom, resentment, and frustration, almost completely devoid of humane alternatives to "the system." To be locked into this system is to feel drained of creative initiative, for the pace is organizationally frenetic and the ante of superficial novelty is constantly being raised. Lacking straightforward cultural coherence, having lost daily intercourse with earthy nature and traditional crafts, our atomized society is constantly agitated by televised violence and vicarious sexuality. Short of economic

II. With the burst "housing bubble," as with its global financial repercussions, it is impossible to avoid marveling at the sincere, straight-faced cynicism of politicians and wealthy businessmen, who otherwise unspeakably detest socialism, rushing to justify the bailout of ("too big to fail") predatory lending institutions with hundreds of billions of federal dollars. What is stunning is not only the brazen hypocrisy but also the amazing speed with which the government is able to mobilize resources when capitalist institutions have cornered themselves in a crisis of manic greed run amok.

breakdown or military holocaust—catastrophes that decide for us—we either remain in the system, based on an increasingly precarious money economy, or we can commit ourselves to working out a way of life that is at once an advance and a retrogression: the renewal of common culture with select appropriate technology. And one of the ways by which to renew common culture is to actively participate in household subsistence.

Regenerative agriculture as household self-provisioning has a long and honorable history. "The Greeks called it *oeconomia*, the etymon of the word 'economy'," writes Karl Polanyi in his excellent work of economic history, *The Great Transformation*:

> Aristotle insists on production for use as against production for gain as the essence of householding proper; yet accessory production for the market need not, he argues, destroy the self-sufficiency of the household as long as the cash crop would otherwise be raised on the farm for sustenance, as cattle or grain; the sale of surpluses need not destroy the basis of householding . . . In denouncing the principle of production for gain 'as not natural to man,' as boundless and limitless, Aristotle was, in effect, aiming at the crucial point, namely the divorcedness of a separate economic motive from the social relations in which these limitations inhered.[3]

The more power this "separate economic motive" gains in atomized society, the more completely do "social relations" deteriorate.

Except in those historic contexts where large estates, owned by nobility and tended by slaves, were common, the record of agricultural production is overwhelmingly characterized by household and village subsistence, even when weighted down by manorial extractions. The emergence of commercial agriculture broke this pattern, both in the corporate agribusiness of the West and in the forced collectivization of the East. Each system, capitalist and communist, forced agriculture into an industrial mode, compelled the acquisition of larger holdings, depopulated (in varying degrees) the countryside, and suppressed rural culture. The promulgation of culture itself became a commercial and bureaucratic undertaking. Participatory folk culture was relegated to the museum and historical archives.

Our task in the wake of this worldwide cultural debacle is to search for the essential roots of folk culture and attempt to rebuild a coherent pattern by which the life of society may be renewed: the military-industrial option to such renewal is too horrible to deserve rational consideration.

The renewal of participatory folk culture requires the enlightened social control of industrial progress.

II

In his essay for *The Futurist*, Robert Rodale asks that we take "an imaginary voyage back to the earliest roots of agriculture and try to see what possibilities for regeneration our ancestors may have missed. Then we can pick up the pieces of history and try to build a series of new and regenerative methods of producing food and fiber."[4] Rodale's primary interest, in this regard, is itself revolutionary: he is advocating a serious investigation into the possibilities of utilizing seed-bearing perennial grasses as a major food source. A shift from annuals to perennials would greatly reduce the need for tillage and radically decrease soil loss due to all forms of erosion. The regenerative capacity of the soil would thereby be indefinitely maintained. Such a regenerative agriculture would be at once a revolutionary advance and a radical retrogression: a synthesis of Carl Jung's proposals for reform. The leading work in the study of perennial grasses is being done by Wes and Dana Jackson at their Land Institute in Kansas, and one can find in their book, *New Roots for Agriculture,* a clear articulation of the problems inherent in conventional agriculture and hopeful, if tentative, indications for the future of regenerative methods.

In advocating and, to some degree, financially supporting such research, Rodale was pointing out that when people "began to cultivate grass plants about 10,000 years ago, they selected the annuals" from the great variety of plant species. Growing annuals required "digging the soil every year, weeding, and finding a way to store seed for next year's planting."[5] This early horticulture was the historic pivot that permitted the emergence of stable villages (stable in location) and enabled human population to increase. Regenerative agriculture has, therefore, cultural as well as methodological implications. We can get a better feel for what this means by examining the historical particulars.

III

In gathering and hunting societies, women were the primary gatherers. It follows from this that women also were the original horticulturalists: from gathering grains to planting them is a short, if also a revolutionary, step. The historian Mary Beard, in her book *On Understanding Women*, contends that women were in fact responsible for domesticating wild grain; she also says that women first raised stock animals, controlled fire, practiced medicine, and took the leading role in spinning and weaving, pottery making, tailoring, and shelter building. The archeologist V. Gordon Childe, in *What Happened in History*, agrees that women grew the grain and made the pots and clothes, but he gives hut building and stock tending over to the initiative of men. In *The Underside of History* by Elise Boulding, and in Lewis Mumford's *The City in History*, one can find a great deal of information on the feminine origins of agriculture, pottery, and, indeed, of village life in general. Elise Boulding says "The earliest development of agriculture, and the settlements around which it was pursued, were women's special work."[6] In a section of her book dealing with mother-goddess themes in Neolithic culture, she asserts:

> With the emphasis on women's fertility, in a society organized primarily on the basis of access to land for farming, control over farm land, farm products, and sheep and goats, as well as control over house, tools, and equipment used by the women, might naturally be passed on through the women.[7]

Mircea Eliade, the well-regarded historian of archaic religions, identifies a great deal of symbolic content in early agriculture. As he points out in *The Sacred and the Profane*:

> It is obvious ... that the symbolisms and cults of Mother Earth, of human and agricultural fertility, of the sacrality of women, and the like, could not develop and constitute a complex religious system except through the discovery of agriculture; it is equally obvious that a preagricultural society, devoted to hunting, could not feel the sacrality of Mother Earth in the same way or with the same intensity.[8]

(Eliade may have misrepresented the "sacrality of Mother Earth," however, in regard to preagricultural hunting. In *The Descent of Woman*, the British anthropologist Elaine Morgan says that outside the colder regions of Earth, women in hunting and gathering societies consistently produced anywhere from sixty to eighty percent of the food. Woman, she says, was the breadwinner; man was the "meatwinner."[9] In a book entitled *People of the Lake*, Richard Leakey—son of the famous team of paleoanthropologists, Louis and Mary Leakey—says that when "meat becomes an important element within a more closely organized economic system so that there exist rules for its distribution, then men already begin to swing the levers of power." He goes on to say that "The status of women sinks further as we go into latitudes which enforce economies based more and more on meat," and finally, among the Eskimos, women "operate in a domestic world in which their social and sexual lives are totally dominated by the wishes of the men."[10]) Aside from these digressive qualifications, however, Eliade nevertheless says flatly that the "sacrality of women depends on the holiness of the earth."[11] Such an assertion is deeply provocative; as an insight, it has tremendous implications for our understanding of women's liberation and environmentalism as social movements. It suggests that an ecological society would also be "feminist." That is, such a society would be far more fully imbued with behavior-shaping values of feminist typology than is our current male-dominated world.

Edward Hyams, in his remarkable book *Soil and Civilization*, also touches into the territory of sexual symbolism:

> As money, cowrie-shells were so firmly established in China as a tradition, that the first metal coins of China were imitations of cowrie-shells in copper . . . [P]rimitive men saw in the cowrie shape and pattern a resemblance to the female vulva, focus of the whole Fertility Cult idea, and therefore of primary and incomparable importance in agriculture.[12]

In *The Transformations of Man*, Lewis Mumford adds more fuel to the feminist fire:

> This domestication of plants seems largely woman's work: instead of taking life, like the hunter, she nurtured it in the earth, as she nurtured it in her own womb . . . With this came hearth and home: a permanent habitation and a regular habit of life, favourable to further nurture . . . The invention of agriculture wrought a further change in man's self-culture; for domestication is a process of gentling and nurturing and breeding that demands selective care and forethought. In every part of this process woman's needs, woman's solicitudes, woman's intimacy and the processes of growth, woman's capacity for tenderness and love, played a dominating part.[13]

One of the crucial factors that undermined this configuration of "gentling and nurturing" was the invention of the plow—that new tool that put men in the fields and, as Childe expresses it, "relieved women of the most exacting drudgery, but deprived them of their monopoly over the cereal crops and the social status that conferred."[14] This rather sud-

den alteration in gender valuation required, quite obviously, the prior do-
mestication of suitable animals for pulling plows: cattle were apparently
domesticated around 6,500 B.C., while the plow has been traced back to
3,000 B.C.

But before the invention and use of plows, again according to Childe,
"owing to the role of women's contributions to the collective economy,
kinship is naturally reckoned in the female line, and the system of 'mother
right' prevails. With stock-breeding, on the contrary, economic and social
influence passes to the males and kinship is patrilinear."[15] Edward Hyams
confirms this idea:

> [W]here masculine economic values dominated, that is in those
> societies where the male business of hunting had given rise to the
> male business of stock-raising, and in due course to nomadic pas-
> toralism, the importance of women declined with the declining
> importance of tillage.[16]

In the words of Elise Boulding, "The shift in the status of the woman
farmer may have happened quite rapidly, once there were two male specializations relating to agriculture: plowing and the care of cattle."[17]

The deliberate breeding of cattle seems to have been a point of departure for patrilineal patterns of descent. William Irwin Thompson, in *The
Time Falling Bodies Take to Light: Mythology, Sexuality, and the Origins
of Culture*, says that once cattle "were kept in corrals, and once second
generations were raised in captivity, then the observation of inherited
characteristics, like Mendel's observations of sweet peas, would encourage a focusing of attention on genetics. It is at the point of conscious stock
breeding that I think one is likely to get the idea of physiological paternity."[18] In other words, the period in which horticulture was discovered, put
into practice, and altered by the "deliberate breeding of cattle," was also
the period characterized by other revolutionary changes in the culture of
social life—perhaps most explicitly the emergence of patrilineal descent
patterns. W. I. Thompson goes on to say:

> Gardening had been women's work; the domestication of animals
> had been the work of man the hunter. When man led his animals
> into the fields for plowing, he surrounded the old culture with a
> new technology. Women's work became a content in a new larger
> structure. When to that relationship of ox and plow was added the
> complex social organization represented by dikes, canals, walls,

storehouses, and recordkeeping, civilization was crystallized out
of the liquid solution of ... village life.

In the movement from village to city the old order of woman
becomes seen as simple and primitive, possessing none of the great
technological wonders of male civilization. A woman is a gardener,
but a man is a serious farmer.[19]

As Thompson also points out:

[W]omen as gatherers had slowly, in the years from 9000 to 6500
B.C., transformed gathering into gardening and agriculture. The
vast storages of grains produced the economic surplus that enabled the men to turn hunting into a nostalgic sport and atavistic
ritual, and trade into their major economic activity ...

The period 6000-4000 B.C. is the *Magnus Annus* of the Neolithic
Great Goddess, but by 4000 B.C. the Near Eastern world is so criss-
crossed with trade routes that the lattice-work itself begins to be a
new cultural entity, the entity of civilization. Civilization involves
not simply the rise of one city, but rather the rise of an urban
nexus in which the specializations of civilization are supportable.
By 4000 B.C. the new world of trade, of craft specializations like
metallurgy and militarism, have created a whole new world, a decidedly masculine world.[20]

"But with the rise of a new economic order through trade and the appearance of skills and specialties associated with the rise of a military class,"
Thompson goes on,

... the social conditions were created in which biological paternity
and the inheritance of private property could become culturally
recognized and socially institutionalized.

It may not have happened overnight, but it was a revolution
nonetheless. For hundreds of thousands of years the culture of
women and women's mysteries had been the dominant ideology
of humanity. The humanization of the primates in the shift from
estrus was a feminine transformation. The rise of a lunar notation
and the beginnings of an observed periodicity upon which all human knowledge is based was a feminine creation. Agriculture and
the rise of sedentary villages and towns were feminine creations.
But civilization and warfare were not; they spelled the end for the
Great Mother.[21]

As we can begin to see, however dimly, the dialectic of advance versus
retrogression contains elements of sexual symbolism in the shifting valua-

tions of gender. The correlations are complex, but they are powerful and real. They provide a great deal of the buried emotional content, unresolved and potent, within our present social arrangements and cultural crises. Our task is to bring those emotional dynamics into open, sympathetic consciousness and begin to work toward an enlightened resolution before the principle of industrial progress and the "specializations of civilization" destroy what's left of their residual existence in agrarian folk culture.

IV

Clearly, civilization developed out of a matrix of forces beyond the domestic stability of settled and well-fed agricultural villages. Leonard Cottrell, for instance, in *The Anvil of Civilization*, lists "the making of bricks for building, the construction of the potter's wheel, wheeled transport, the sailing ship, and the harnessing of domestic animals for transport and haulage" as some of the achievements that gave rise to civilization. He goes on to say that the "most important discovery was the metallurgy of copper and bronze."[22] It seems accurate to recognize in these advances the growing specializations of men (hunting having become, in Thompson's words, an "atavistic ritual"), while the actual world of women was being steadily compressed. (Cottrell unintentionally reinforces this view when he talks repeatedly of men and "their womenfolk"—as if women were mere possessions, much like cattle.[23])

According to Elise Boulding, "It seems clear that urbanization in many ways 'encloses' women."[24] Although Lewis Mumford says that the city is the "most precious collective invention of civilization . . . second only to language itself in the transmission of culture," he also suggests that the "most important agent in effecting the change from a decentralized village economy to a highly organized urban economy was . . . the institution of Kingship," and that this change was "accompanied by a collective shift from the rites of fertility to the wider cult of physical power."[25] This shift from female fertility to male power can only imply that feminine influence, in all its various social and cultural manifestations, was declining.

The city, then, though its vital nucleus was the feminine agricultural village, rapidly developed as a fortress of specialization, trade, and war whose institutional structures were—and are—controlled by men. The feminine village became the "city of man." The city maintained its existence by aristocratic regimes that expropriated the labor of the village and the bounty of Earth, enlarging their empires by means of war. Our present global crisis is, to a very large degree, the historic unfolding of these predatory institutions in an industrial mode. Civilized predation has become "democratized." For this depth of crisis, there is no technical fix. We will survive by spiritual humility and cultural transformation, or we will not survive at all.

ENDNOTES

1. Methvin, "TV," 50.
2. Mander, *In the Absence*, 76.
3. Polanyi, *Great*, 53–54.
4. Rodale, "Breaking," 18.
5. Rodale, "Breaking," 18.
6. Boulding, *Underside*, 112.
7. Boulding, *Underside*, 141.
8. Eliade, *Sacred*, 17.
9. Morgan, *Descent*, 160-61.
10. Leakey, *People*, 248.
11. Eliade, *Sacred*, 144.
12. Hyams, *Soil*, 158.
13. Mumford, *Transformations*, 26–27.
14. Childe, *What*, 13.
15. Childe, *What*, 66.
16. Hyams, *Soil*, 286.
17. Boulding, *Underside*, 163.
18. Thompson, *Time*, 127.
19. Thompson, *Time*, 164.
20. Thompson, *Time*, 150, 155–56.
21. Thompson, *Time*, 156.
22. Cottrell, *Anvil*, 14.
23. Cottrell, *Anvil*, 13.
24. Boulding, *Underside*, 191.
25. Mumford, *City*, 53, 35, 39.

4

Rivals to the Hearth

ACCORDING TO THE LATE economist Ralph Borsodi, a woman's place was in the home. For him, the demise of the self-provisioning household, in which a woman's economic contribution equaled if not exceeded that of a man's, was a loss of unprecedented magnitude. He was a lifelong foe of the industrial system, and an active partisan of decentralized rural life. One might call him a Jeffersonian-with-electricity, for he believed in all sorts of tools and gadgetry (all functional, to be sure) that lightened the work of the household. I think Borsodi was not at all a raw sexist but, rather, a person who associated hearth with heart, and to both he gave feminine attributes.

Yet in this day of assertive feminism, when women are claiming as a basic human right full participation in all facets of the economy and at all levels of social and political life, Borsodi's views on the proper place of women no longer earn him the good housekeeping stamp of approval. Gender-specific work is being challenged, one might safely say, as never before in human history. Traditional household economy has been ruined. It is indeed possible that gender-specific work patterns and sex-based stereotypes are being broken through in such a way and to such an extent that gender patterns will never again regain the exclusive power they once contained. Or, where older models are maintained, there will be an element of voluntarism associated with them that was unknown to the past.

Yet it is true that the very uncertainty introduced into the self-provisioning household by replacing tradition with voluntarism (our context here is the nuclear family) makes the modern household much more vulnerable to partial incompetence or total disintegration. This new instability brings into question the viability of the self-provisioning nuclear homestead itself as a reliable or practical means of reform by retrogression; if tasks can no longer be determined by gender, and if all tasks assume some degree of voluntarism, then there is no longer any guarantee that all the tasks necessary for self-provisioning will in fact get done.

For those who recognize or who have experienced this problem personally, who cherish not only the personal rewards arising from self-provisioning but who also insist that there is wide-spread social stability to be gained by extensive participation in householding, the dilemma generated by the conflict between voluntarism and necessity can become intense. To embrace necessity as primary implies a loss of voluntarism—a loss that many people, especially feminists, are not likely to accept. To hold voluntarism as primary leaves the door open to inadequate household provisioning and can provoke an anxiety that may lead to disruption. Let us look briefly into the history of this dilemma.

II

In the wreckage of preindustrial folk culture, traditional patterns of folk subsistence have been lost. The evidence is great, however, that, despite a rough equality between women and men in the actual practice of household provisioning, it was the men who held power in the legal and social apparatus, in the public sphere. When it comes to the issue of power, our history focuses on wars, armies, knights, generals, kings, and popes—and all this has been (and mostly remains) the playground of men. Furthermore, to compound the power differential in regard to gender, there was a chasm between the upper class and the working class, between the aristocratic rulers and primary producers. The Industrial Revolution pitted the relatively unambitious working class—people largely from the countryside who carried a cultural *tradition* of subsistence and sufficiency—against the sharp commercial ambitions of a rising urban middle class in alliance with elements of the landed aristocracy. (There were, however, members of the old aristocracy who detested the new industrial conditions and their entrepreneurial promoters.)

The crushing of folk culture between the stones of mass production and the enclosure of the commons created a cultural vacuum. Where once there had been vital peasant culture, there was now a stricken, demoralized proletariat. Into this cultural vacuum flooded urban middle-class rigidity, replete with uniform values: nonseasonal regularity, stifling routine, punctuality, obedience to managerial authority, cleanliness, sobriety, religious revivalism, chastity, and piety. All this is part of what E. P. Thompson, in his voluminous book *The Making of the English Working Class*, calls the "psychic ordeal in which the character-structure of the rebellious pre-industrial labourer or artisan was violently recast into that of the submissive industrial worker."[1] This earlier rebelliousness presupposes a depth of culture, and the "character-structures" arising from that culture, that were inherent in preindustrial life. It presupposes an economic *independence* of the folk community, a collective capacity for self-provisioning that not only suggests the lower classes did not economically *need* the upper class, but that they saw the upper class as an economic parasite. But with enclosure, the necessities of life increasingly had to be obtained in the marketplace as commons were destroyed and guilds undermined. Folk suspicion of the emerging commodity utopia was broken down by the enforced extinction of those social conditions that enabled folk culture to exist; the reliance on self-provisioning at the level of household and village community is the key to the stability of folk culture. Crush self-provisioning and folk culture is doomed. The eviction from the commons, in England, guaranteed this crushing.

The building of an enlightened and democratic society, in which stable forms of self-provisioning are readily available and widely practiced, is obviously a major political and cultural undertaking. Such a project flies in the face of the industrial ideology that has sought to eradicate not only the entrenched affection for subsistent living, but the possibility of subsistence itself—just as the enclosures and evictions in early industrial England forced country people either onto starvation parish relief or into starvation piece-rate employment in the hated factories. This industrial ideology is now so pervasive and so entrenched that its detractors are seen as immature if not irrational. It is hardly an overstatement to say that those who oppose unchecked industrial progress and unlimited economic growth are considered by industrial apologists as either primitive and deluded or backward and subversive. But nothing merits deeper scrutiny than that which nearly everyone takes for granted. We are captivated by civilization, its economic system and its progressive ideology, because the culture of the ecological folk commons has been destroyed and its disappearance celebrated as a blessing, as a civilized advance. With enforced industrialization, civilization goes beyond the mere impoundment of folk culture and enters the realm of cultural extermination.

Precisely because of Ralph Borsodi's views on the role of women in the household economy, he is a valuable source of information on the relationship between the factory system and the women's movement. In his book *This Ugly Civilization*, Borsodi says that "In changing the economic foundation of the family from a domestic production to a factory production basis, the factory changed the entire social status of women and children." The Industrial Revolution, he says, "created a host of novel political, legal, economic and social problems: child labor, trade unionism, universal manhood suffrage, socialism, woman suffrage, economic and sex independence for both men and women." He goes on to say that "In addition to credit for its contribution to creature comfort, the factory has to be given credit for increasing the political freedom of men and women."[2]

One must understand, however, that the factory system did not bring either of these advances—creature comfort or political freedom—to workers in any spirit of philanthropic benevolence. It was in reaction *against* the factory system, and *against* the utter ruthlessness of factory owners and their financial and ideological allies, that each gain in freedom and well-being was won by working people in grim and even deadly political battles. The factory system simply wrecked the preindustrial social structure. The tragic irony regarding the gains in freedom and well-being that have since been won is that these gains have only been practicable *within* the industrial system. And while these industrial gains were consolidated in the twentieth century (before economic globalization and rampant "free trade" began to undo the workers' relative security and affluence), the memory of the commons was largely lost. The irony of unionization in America is that it tended to result in an increasingly close identification, not with folk solidarity, but with nationalism and with American foreign-policy adventures, as in Vietnam, where rice-paddy peasants, shot by the sons of the working class, became victims of economic and political aggression. With "free trade," unions were broken even as the working class began to spurn "liberalism" and align itself with "conservative" politicians like Ronald Reagan, who opened his presidency by busting a union. Worker disorientation was both tragic and epidemic. This disorientation was the consequence of the working class losing the "character-structure" of preindustrial culture—the loss of folk grounding, in other words, without a corresponding gain in the insights and convictions of what we might call ecological socialism. Into this vacuum flooded conventional consumerism and reflexive nationalism.

E. J. Hobsbawm, in his *Industry and Empire*, says that the "traditional world and way of life" of the laboring poor was "destroyed, without automatically substituting anything else. It is this disruption which is at the heart of the question about the social effects of industrialization."[3] E. P. Thompson states the workers' attitude toward the factories more explicitly:

> It is easy to forget how evil a reputation the new cotton mills had acquired. They were centres of exploitation, monstrous prisons in which children were confined, centres of immorality and of industrial conflict; above all, they reduced the industrious artisan to 'a dependent State.' A way of life was at stake for the community, and, hence, we must see the croppers' opposition to particular machines as being much more than a particular group of skilled workers defending their own livelihood. These machines symbolized the encroachment of the factory *system*. So strongly were the moral presuppositions of some clothiers engaged, that we know of cases where they deliberately suppressed labour-saving inventions . . .[4]

Karl Polanyi, in *The Great Transformation*, says that during the early days of industrialization, the capitalists, both in Europe and in Europe's colonies, practiced the "smashing up of social structures in order to extract the element of labor from them." He says that "the early laborer, too, abhorred the factory" and that "it was necessary to liquidate organic society, which refused to permit the individual to starve."[5] It is in this light that we must understand Borsodi's remarks on the connection between the factory system and individual freedom. Polanyi elaborates on the process:

> Hobbes' grotesque vision of the State—a human Leviathan whose vast body was made up of an infinite number of human bodies— was dwarfed by the Ricardian construct of the labor market: a flow of human lives the supply of which was regulated by the amount of food put at their disposal. Although it was acknowledged that there existed a customary standard below which no laborer's wages could sink, this limitation also was thought to become effective only if the laborer was reduced to the choice of being left without food or of offering his labor in the market for the price it would fetch. This explains, incidentally, an otherwise inexplicable omission of the classical economists, namely, why only the penalty of starvation, not also the allurement of high wages, was deemed capable of creating a functioning labor market.[6]

The factory was the means by which "organic society" was destroyed, and out of that social chaos emerged the various movements and political causes that Borsodi enumerates.

Mary Beard, commenting on the same period, says "the family, transferred from the country to the city, encountered rival interests to the association of the hearth and was pulled apart—a movement involving a disruption of the man and wife relation and the parent-child authority."[7] Referring to the power loom, steam engine, locomotive, electric light, and automobile, and giving the names Cartwright, Arkwright, Crompton, Watt, Stephenson, Edison, and Ford, she says:

> Such men were the agents, guileless about ultimate consequences, of a radical change in the modes of caring for life which removed them further from feminine control than had ever been the case before and in fact made man director of physical comfort for the first time.[8]

Beard goes on to say that in the "initial stages of the industrial revolution, at least, women lost heavily in economic independence and prestige," and, from that point on, the "story of industrial change is in part the individualist account of wives and daughters struggling to regain what they had lost in the form of economic independence."[9] She also says:

> If the established economic mill keeps running, then the future before various women is clear. They will acquire through inheritance, transfer, or their own efforts more property with all the rights and privileges thereunto attached. The number of professions and employments open to women will increase.[10]

Mary Beard's predictions (*On Understanding Women* was published in 1931) have certainly come true—up to a point, and largely in terms of individualistic success. There has been, and still is, a women's movement. Many more women have gained access to the professions. We also have in the United States what has come to be called the "feminization of poverty," as those women, especially single mothers, who are not established in the system are reduced to a socially degraded dependency. Along with this, there is in the present economic dispensation a vast increase in reports of battery and abuse.

"In two periods," writes Shepard B. Clough in *The Rise and Fall of Civilization*, "in the millennium prior to 3000 B.C. and in the four and a half centuries since 1500, the development of new techniques was crucial in economic progress."[11] Yet the emergence of these "new techniques" was closely associated, in both instances, with factors other than pure economic progress; in both cases we can trace a *dissociation* of life around the feminine hearth and new levels of masculine control over what Mary Beard calls the "modes of caring for life." In both periods, women *lost* important areas of economic independence and social prestige; in both instances, specializing men *gained*. The developmental stages in the creation of an urban-industrial world have greatly increased the overall power of men. The drive to "conquer nature" has reached, with nuclear weaponry, apocalyptic proportions. Men, with their drive for control and domination, have brought us to the foot of the global gallows. Or, perhaps we might also say that the nearly simultaneous industrialization and globalization of civilization has brought us to an ecological and cultural precipice. Now what?

III

In considering options for the renewal of self-provisioning society in the context of a democratic and ecological culture, there are three patterns of subsistent living that come easily to mind. The first model is that of a single person who lives the austere life of, say, Henry David Thoreau. This is not a pattern capable, quite obviously, of reproductive renewal. Nor should we overlook the fact that the short-term hermit of Walden Pond shared many a meal with his aunt in nearby Concord and sat often at Emerson's table. Yet Thoreau stood aside from the society of his day precisely because it seemed to him so devoid of sharing and community. He was an individualist only in a limited sense; his interest and need lay in balancing solitude with a kind and quality of community that wasn't readily available to him and that probably didn't exist to any appreciable extent in his day, despite the attempts at intentional community (like Brook Farm) going on in Massachusetts in the 1840s and 1850s.

The second option for the renewal of self-provisioning is that of the couple or family. It is the nuclear family with which we are most familiar and that, by being nuclearized, has suffered many of the heaviest losses in the industrial war on subsistence. The breakdown of gender-specific work has rendered the nuclear family—itself a creation of the factory system—a questionable agency for the renewal of self-provisioning society. What has been called the "new homesteading movement" is caught squarely in these contradictions.

While it is true that "The new homesteading movement provides alternatives to urban life and its sense of dispossession and meaningless labor, to the 'need' to consume artificial goods and inferior products, to the ideology of progress and futurism, and to an economy based on the idea of scarcity," as Maynard Kaufman insists in his essay "The New Homesteading Movement" (published in the Spring 1972 issue of *Soundings*), it is not surprising that those attempts at homesteading that have most fully succeeded are those that practice gender-specific work patterns.[12] It is also not surprising that so few people have chosen to enter into homesteading on those terms—which is one reason why the new homesteading "movement" remains a fairly small social phenomenon.

The third pattern is one with which we have the least experience and, perhaps, the greatest anxiety: cooperative subsistence. Our tendency in this regard is to think loosely, and no doubt pejoratively, of "communes," with all the negative connotations associated with the instability of hippie experiments, or, alternatively, with the rigidity of religious-based communities, from the Amish to the monastery. As an industrial society, we have all but forgotten that traditional folk culture was heavily shaped by, and existed around, the village commons. The enclosure acts struck at the very root of folk culture: without a common ground there could be no common culture. The private family farm has shown itself incapable of sustaining the continuity of rural culture in the face of the commercial and industrial economic agenda. So it is here, in the territory of cooperation, where we must make a new and serious appraisal. And we would do well to make that appraisal from within a context that Ralph Borsodi helped define: the land trust.

The land trust is a highly flexible legal concept. It is, according to The Institute for Community Economics of Springfield, Massachusetts, "a democratically structured non-profit corporation, with an open membership, created to hold land permanently for the benefit of a local community." The land trust is a way of "owning" land in common, but with a great range of options for leaseholds and inheritance. Rightly utilized, the land trust could well provide the reality to Aldo Leopold's proposal that we learn to live as community on the land. In the Foreword to *A Sand County Almanac*, Leopold said "We abuse land because we regard it as a commodity belonging to us. When we see land as a community to which we belong, we may begin to use it with love and respect."[13]

A community is vulnerable to disruption to the degree that it is incapable of preventing undesirable use or exploitation of its immediate physical environment. (This is one way of describing the fate of American

Indians under the impact of intrusive European civilization. They were simply overpowered. If war is only a continuation of diplomacy by other means, as the Prussian general Karl von Clausewitz said, then the destruction of nonindustrial cultures is only another form of this "diplomacy.") But by taking land off the auction block of commodity speculation, by formulating principles of use that guarantee sound environmental practices, and by practicing the arts and crafts of economic self-provisioning, a community land trust could build toward ecological stability and communal coherence. To what extent, or in what way, a land trust would be a functional cooperative is of necessity the consensus of each "cooperative." The range of options—from lifetime, renewable leaseholds to commons in perpetuity—is unlimited.

We must understand, however, that only a community acting *as a community* is capable of protecting the land in any long-term way—and even then only insofar as it is capable of preserving its political existence. When we consider the cultural debacle brought on by industrialization, it's important to see to what extent preindustrial society was immersed in practices of mutual aid. We must remember that private property as we understand it—that is, privately held "real estate" bought and sold as a commodity, hedged as equity, and available to any buyer with the requisite financial assets—simply did not exist to any appreciable extent before the Industrial Revolution. Except for a few yeoman farmers who held title to small parcels of land, the secular aristocracy and the church controlled virtually all the land in Europe. The folk community lived in villages and farmed the commons and the manorial fields; it was the subsequent eviction of the peasants from the commons, for the purpose of commercial agriculture at the behest of the upper class and aspiring middle-class entrepreneurs, that the initial steps in folk culture devastation took shape. As E. P. Thompson puts it:

> In village after village, enclosure destroyed the scratch-as-scratch-can subsistence economy of the poor—the cow or geese, fuel from the common, gleanings, and all the rest. The cottager without legal proof of rights was rarely compensated. The cottager who was able to establish his claim was left with a parcel of land inadequate for subsistence and a disproportionate share of the very high enclosure costs . . . It became a matter of public-spirited policy for the gentleman to remove cottagers from the commons, reduce his labourers to dependence, pare away at supplementary earnings, drive out the smallholder.[14]

In the above passage, Thompson is writing of the English experience in the late eighteenth and early nineteenth centuries. But the industrial attitude toward folk culture and subsistence livelihood is no less fierce now than it was two hundred years ago; it has simply become normative.

The cultural as well as the methodological implications of regenerative agriculture must be emphasized. In a world where political and economic rivalries have become so intense that we live always under the cloud of annihilation, and where the instruments of destruction available to the civilized international adversaries are so cataclysmic and ecocidal, we are in urgent need of social policies that address and reduce the discontents inherent in industrial civilization. As Maynard Kaufman says in his essay on homesteading:

> It is not only possible but desirable that the future should be enriched with things of value and ways of life kept from the past. And if it is necessary to go 'back' to nature to establish our relationship to it, let us by all means go 'back.' What use is our awareness of history if it does not free us from the tyranny of historical determinism or from continual and impoverishing change? And what good is affluence if it means that nothing can be treasured?[15]

As industrial technology constantly seeks to raise the ante of progress and growth, it contributes, perhaps unwittingly, to a rise in international hostility.[1] It is sheer madness to reject thoughtful retrogression out

---

1. There are commentators who believe the sort of "terrorism" associated with the attacks on September 11, 2001, is, in fact, a kind of "blowback," a violent refusal by those who felt pushed around or humiliated to accept relentless imperial imposition from the West. Those attacks—two commercial jetliners flown into the Twin Towers in New York City, another hitting the Pentagon, and a fourth crashing into a field in Pennsylvania—perhaps hardly more than audacious slaps in the face, but stunning nevertheless—were enough, however, to instantly align public will behind the federal government's desire for retaliation and revenge. In this supposedly Christian nation, there was no talk of exploring motives, turning the other cheek, or loving the enemy. That is, in the righteous indignation of aggrieved victimhood, there was no room for humbly examining whether the attackers might have had, at some level, legitimate grievance on their side, too. To even suggest such a possibility verged on treason. But here we arrive at a watershed. If a Christian ethics is not permitted to shape "Christian nation" policy, then we are compelled by spiritual evasion to align ourselves with raw and brutal retaliation, even as this retaliation dresses itself in righteous victimhood. What is at stake here is not merely a typical and conventional human trait towards self-preservation and self-justification, but the refusal of a privileged economic system to acknowledge its disproportionate advantages and its unwillingness to relinquish those advantages through global humanitarian sharing within ecological limitation. Thus a future rife with increasing fear, surveillance,

of arrogant "sophistication"; such rejections move us relentlessly toward even more probable destruction.

Industrialism managed to create and sustain an ideological hegemony for at least two reasons. First, it destroyed the solidarity of the folk community and the means of its existence; second, it substituted a blind affirmation in technical progress for the traditional suspicion of unknown and uncontrolled change. This process turned the technical expert into a new priestly authority whose educated (or glib) salesmanship substituted for clerical authority. If the power of the church rested, theologically speaking, in the promise of an eternal afterlife of bliss, then the power of the industrial economy was, more immediately and alluringly, contained in the mass mirage of Progress. If the church had individualized the spiritual future in the direction of private salvation, then the industrial economy individualized the economic future in a way unique to human history. The economist Robert L. Heilbroner, in *The Worldly Philosophers*, provides us with a larger historical perspective:

> The profit motive as we know it is only as old as 'modern man.' Even today the notion of gain for gain's sake is foreign to a large portion of the world's population, and it has been conspicuous by its absence over most of recorded history . . . The idea of gain, the idea that each man not only may but should constantly strive to better his material lot, is an idea which was quite foreign to the great lower and middle strata of Egyptian, Greek, Roman, and medieval cultures, only scattered throughout Renaissance and Reformation times, and largely absent in the majority of Eastern civilizations . . . The absence of the idea of gain as a normal guide for daily life—in fact the positive disrepute with which the idea was held by the church—constituted on enormous difference between the strange world of the tenth to sixteenth centuries and the world that began, a century or two before Adam Smith, to resemble our own. But there was an even more fundamental difference. The idea of 'making a living' had not yet come into being. Economic life and social life were one and the same thing . . . The Middle Ages, the Renaissance, the Reformation—indeed the whole world until the sixteenth and seventeenth centuries—could not envisage the market system for the thoroughly sound reason that Land, Labor, and Capital—the basic agents of production which the market system allocates—did not yet exist. Land, labor and capital in the sense of soil, human beings, and tools are of course coexistent with

---

restrictions, and violence is assured. This is where the doctrine of "two kingdoms" leads.

society itself. But the idea of abstract land or abstract labor did not immediately suggest itself to the human mind, any more than did the idea of abstract energy or matter. Land, labor, and capital as 'agents' of production, as impersonal, dehumanized economic entities, are as much modern conceptions as the calculus. Indeed, they are not much older.[16]

What the new industrial process engendered, says Heilbroner, was nothing less than the "commercialization of life."

It is this materialistic motive that marks the crucial divide between radically different ways of living: one ancient, destroyed, all but forgotten; the other new, expansive, all pervasive. It now remains to be seen whether traditional householding, enacted in a newly democratic and cooperative manner, is indeed recoverable to people attracted to the possibility of living in a renewed common culture. Yet because we have come to be so fiercely protective of our individualized autonomy, we fear a loss of freedom through cooperative association. We have forgotten how to live "in common." We have, by virtue of capitalist propaganda, become frightened of all words that derive from the Latin *communitas*. Whether we can learn to face our fear and examine it, whether we can muster the moral courage toward building a new community life, is what remains to be seen. Given the depth and breadth of our social, economic and political anxieties, the rediscovery of community is the great urgency of our time.

ENDNOTES

1. Thompson, *Making*, 367–68.
2. Borsodi, *This*, 133, 67, 76.
3. Hobsbawm, *Industry*, 84.
4. Thompson, *Making*, 548.
5. Polanyi, *Great*, 164, 165.
6. Polanyi, *Great*, 164.
7. Beard, *Understanding*, 512.
8. Beard, *Understanding*, 499.
9. Beard, *Understanding*, 505.
10. Beard, *Understanding*, 518.
11. Clough, *Rise*, 260.
12. Kaufman, "New," 71.
13. Leopold, *Sand*, xviii–xix.
14. Thompson, *Making*, 217, 219.
15. Kaufman, "New," 75.
16. Heilbroner, *Worldly*, 11–15.

5

Loose Aggregations

S O FAR THIS ANALYSIS may seem negative in the extreme: a grim pointing toward the undesirable traits within compulsive industrialization,
with land trusts and rural villages as wishful alternatives to industrial and
technological gloom. Yet even if one is disinclined to accept the probability of serious breakdown or dislocation—though such dislocations are
the episodic norm within civilized history—it remains true that society as
presently construed simply does not provide a rich cultural life or a vital
human community. For the more clever, better educated, or more highly
ambitious, society offers a slick professional careerism. But safe and well-
padded careers, too, are no substitute for cultural vitality and community
coherence; such careerism has become part of the erosion of those social
cohesions that, in the past, handled many problems in the community
rather than passing them off to experts for resolution.

This is not to say that traditional community always acted wisely or
well. But neither do the social service agencies that have replaced community always act wisely or well, even with the loftiest of intentions. If
there is to be deliberate, intelligent, and thoughtful human action aimed
at the creation—mysterious and unmechanical as that may be—of stable
human community, it has to be undertaken with the foreknowledge that
true community is not made up of fragmented associations between
people (such as we find in bureaucracies) but rather in long-term com-
mitments to both persons and place. Because cities have become so
commercially bloated, and because the cultural ties between cities and
the countryside are so tenuous, it is virtually impossible to avoid being
smothered by the support and delivery systems of the city, by its mood
of rootlessness, commercial sterility, and violence.[1] The city has become

---

1. By violence I mean not only the violence of urban criminality and street gangs,

a commercial hothouse that boils away even the residues of traditional urban amenities—the small cafes, the street peddlers, the accessibility of parks. The commercial city towers over the countryside in a way directly analogous to how the wealthy have risen above the poor. A true community must have roots in the natural world; it must strictly limit the accumulation of wealth and power in the hands of a few.

Independence in an affluent society afloat on electronic images of effortless allurement has translated into a cynical aloofness from social conflict and a derisive contempt for all those who lack money or style. Multinational business now exploits, whenever and wherever possible, the less expensive labor of the Third World poor. Therefore unionized American workers find their wages pulled down by Third World "competition." These American workers are pressured to accept lower wages— "free trade" has been facilitated by deunionization—while corporations and financial institutions concentrate incredible wealth. Such inequality has grown into a conceited, overblown sense of entitlement for those who run the system. It has become parasitic and decadent, degenerating into a highly stylized consumer neurosis that demands pampering and becomes preoccupied with roles, scripts, self-concepts, emotional contracts, and a cheap therapeutic ethic that justifies limitless self-assertion.

It is simply not possible for any atomized society, sustaining itself by means of a money economy, to develop a stable and resilient community life. Organization begins to substitute for culture, business success for communal vitality. But it is in the common tasks of life, in the "economy" of older meaning, that the mood and vitality of daily life are sustained. In the sharing of common tasks, a quality of connectedness develops among people that can grow in no other way. Combined with spiritual affinity, a shared work life provides a powerfully coherent solidarity. True culture is found in the nonbureaucratic cooperation of daily life. This is the promise that lies beyond the fear of freedom's loss in cooperative association.

For people whose entire economic life is structured by industrial organization or corporate money, there can be no sustained community.

for these can to some extent be explained as the inevitable price of long-term racial discrimination and economic disempowerment, but the brutal violence made manifest by the incredible magnitude of local, state, and federal incarceration, by the proliferation of SWAT teams, and by a military budget larger than the rest of the world's combined, a budget that directly reflects the determination of corporate America to be the permanent bully on the planetary block. Such violence is increasingly built into the very institutions that lock corporations and governments into a revolving-door carousel.

(Labor unions that have concentrated solely on wages and fringe benefits as their reason for existence are now relatively easy pickings for corporate power. They lack communal solidarity precisely because they never developed an alternative social vision. Having barely survived the crushing of their peasant origins, and having been politically frightened into a harsh repudiation of socialism, industrial workers have often reverted to reflexive scapegoating, angry and willing participants in blaming liberals, blacks, Hispanics, feminists, homosexuals, and Muslims for virtually all the troubles in American society. As their middle-class advantages contracted, workers and their shrinking unions had no deeper analysis to fall back on. They were plucked like ripe fruit as "Reagan Democrats.") Economics must take on a more historic meaning: cooperative action and common work for a communal subsistence. When money forms the boundary and limitation of human interaction, there can be no deeper connectedness. When everyone lives off abstract money, and off the economic system that money lubricates, transfuses, and breeds, it is as if there exists an invisible grid that separates each from all. Culture becomes an electronic pastime.

Just as in a modern subdivision where no commons exists, and where each household is linked by private driveway to the public thoroughfare, so too are individuals and families tied into the market economy and dependent upon it for their very survival—each person hooked into bureaucratic institutions as tightly as the garage is tied to the street. Without real community, each family remains dependent on the institutional pay check, the supermarket, the school system, the credit card, the insurance policy, and the police department. Many people apparently desire this state of affairs; they call it independence. But this institutionally dependent individualistic servitude creates the conditions in which mass political manipulation becomes easy, possible, and perhaps inevitable. This kind of "individualism" emerges as an economic consequence and political abstraction from the mutilation of organic society. When people live off the institutional pay check, money transmogrifies into food, clothing, shelter, entertainment, and the general glut of consumer goods; money stands between a person and destitution. But in community, it is real productive skills and specific other people who provide the social "safety net." This distinction makes all the *cultural* difference in the world.

II

Before getting more deeply into issues involving rural culture, it may be necessary to clarify and elaborate some related concepts, especially steps that lead sequentially from the individual to the family to the community to society and, finally, to the state.

Despite the glorification of great individuals in our history books, the basic reality in human history has been the group, of which the most basic pattern, however structured, is the family. We are now living in a period in which the family—the extended family and the nuclear family—has come under severe pressure: divorce statistics, if nothing else, show this to be true. While practically all forms of work and production have come under the uniformity of the machine and factory revolutions, and children under laws of compulsory schooling are forced into educational incompetence, the state has grown in strength and size simultaneously with the decay of local and regional identities. The family, that in preindustrial society had a spreading network of roots like any healthy and native weed, is now bounded by the word "nuclear," as if it were a potted hybrid, obedient before industrial ideology as its predecessor was fearful of the church. Frustrated with the narrowness of the modern family, and unable to remain permanently distracted by the electronic circus of the mass media, many people have taken the next logical step by embracing the individualistic goals of psychotherapy—to become "assertive," to become one's "real self," to maximize "personal growth," and so on. It is true that various forms of psychotherapy have provided real help to many people, including a great many women who have needed to make contact with a dormant emphasis of self in order to break out of the passiveness and deference associated with the "feminine mystique." It is also true, however, that the so-called human potential movement has been largely contained within a middle-class world dependent on the institutional system: to become "assertive" has typically implied an effort to rise in the system rather than make do outside of it. This is the present internal conflict within the women's movement: whether to join the predatory system or create a new ecological culture.

Society, the collective patterns of organic relationships outside the regimentation of mechanical production or the bureaucratic state, was crushed as the "modernization" of society took shape. Work and economic survival became simultaneously abstract and privatized. Society has become merely "loose aggregations of individuals, connected by no particular bonds." So said Peter Kropotkin in his turn-of-the-century book, *Mutual Aid*. Although Kropotkin's works were influenced, inevitably, by the anthropological conceptions of his time, he nevertheless was one of the first articulate opponents of social Darwinism, the doctrine of dog-eat-dog progress that came out of nineteenth-century economic theory meshed with a brutal misreading of Darwin's theory of evolution. But let us indulge in a lengthy passage from Kropotkin's important book, for in it he spells out in clear and vivid terms the conflict between organic society and the state, to the result of an indifferent and withdrawn individualism:

> After having passed through the . . . tribe, and next through the village community, the Europeans came to work out in medieval times a new form of organization, which had the advantage of allowing great latitude for individual initiative, while it largely responded at the same time to man's need for mutual support. A federation of village communities, covered by a network of guilds and fraternities, was called into existence in the medieval cities. The immense results achieved under this form of union—in well-being for all, in industries, art, science, and commerce—were discussed at some length in two preceding chapters, and an attempt was also made to show why, towards the end of the fifteenth century, the medieval republics—surrounded by domains of hostile feudal lords, unable to free the peasants from servitude, and gradually corrupted by ideas of Roman Caesarism—were doomed to become a prey to the growing States.
>
> However, before submitting for the three centuries to come, to the all-absorbing authority of the State, the masses of people made a formidable attempt at reconstructing society on the old basis of mutual aid and support. It is well known by this time that the great movement of the reform was not a mere revolt against the abuses of the Catholic Church. It has its constructive ideal as well, and that ideal was life in free, brotherly communities. Those of the early writings and sermons of the period which found most response with the masses were imbued with ideas of the economical and social brotherhood of mankind. The 'Twelve Articles' and similar professions of faith, which were circulated among the German and Swiss peasants and artisans, maintained not only everyone's right to interpret the Bible according to his own understanding, but also included the demand of communal lands being restored to the village communities and feudal servitudes being abolished, and they always alluded to the 'true' faith—a faith of brotherhood. At

the same time scores of thousands of men and women joined the communist fraternities of Moravia, giving them all their fortune and living in numerous and prosperous settlements constructed upon the principles of communism. Only wholesale massacres by the thousand could put a stop to the widely spread popular movement, and it was by the sword, the fire, and the rack that the young States secured their first and decisive victory over the masses of the people.

For the next three centuries the States, both on the Continent and in these islands, systematically weeded out all institutions in which the mutual-aid tendency had formerly found its expression. The village communities were bereft of their folkmotes, their courts and independent administration; their lands were confiscated. The guilds were spoliated of their possessions and liberties, and placed under the control, the fancy, and the bribery of the State's official. The cities were divested of their sovereignty, and the very springs of their inner life—the folkmote, the elected justices and administration, the sovereign parish and the sovereign guild—were annihilated; the State's functionary took possession of every link of what formerly was an organic whole. Under that fatal policy and the wars it engendered, whole regions, once populous and wealthy, were laid bare; rich cities became insignificant boroughs; the very roads which connected them with other cities became impracticable. Industry, art and knowledge fell into decay. Political education, science, and law were rendered subservient to the idea of State centralization. It was taught in the Universities and from the pulpit that the institutions in which men formerly used to embody their needs of mutual support could not be tolerated in a properly organized State; that the State alone could represent the bonds of union between its subjects; that federalism and 'particularism' were the enemies of progress, and the State was the only proper initiator of further development. By the end of the last century the kings on the Continent, the Parliament in these isles, and the revolutionary convention in France, although they were at war with each other, agreed in asserting that no separate unions between citizens must exist within the State; that hard labor and death were the only suitable punishments to workers who dared to enter into 'coalitions.' 'No State within the State!' The State alone, and the State's Church, must take care of matters of general interest, while the subjects must represent loose aggregations of individuals, connected by no particular bonds, bound to appeal to the Government each time they felt a common need. Up to the middle of this century, this was the theory and practice in Europe. Even commercial and industrial societies were looked at with suspicion. As to the workers, their unions were treated as unlawful almost within our own lifetime in this country [England] and within the last twenty years on the Continent. The whole system of our State education was such that up to the present time, even in this country, a notable portion of society would treat as a revolutionary measure the concession of such rights as everyone, freeman or serf, exercised five hundred years ago in the village folkmote, the guild, the parish, and the city.

The absorption of all social function by the State necessarily favoured the development of an unbridled, narrow-minded individualism. In proportion as the obligation towards the State grew, the citizens were evidently relieved from their obligations towards each other. In the guild—and in medieval times every man belonged to some guild or fraternity—two 'brothers' were bound to watch in turns a brother who had fallen ill; it would be sufficient now to give one's neighbor the address of the nearest paupers' hospital. In barbarian society, to assist at a fight between two men, arisen from a quarrel, and not to prevent it from taking a fatal issue, meant to be oneself treated as a murderer; but under the theory of the all-protecting State the bystander need not intrude; it is the policeman's business to interfere, or not. And while in a savage land, among the Hottentots, it would be scandalous to eat without having loudly called out thrice whether there is not somebody wanting to share the food, all that a respectable citizen has to do now is to pay the poor tax and to let the starving starve. The result is, that the theory that maintains that men can, and must, seek their own happiness in a disregard of other people's wants is now triumphant all around—in law, in science, in religion. It is the religion of the day, and to doubt its efficacy is to be a dangerous Utopian. Science loudly proclaims that the struggle of each against all is the leading principle of nature, and of human societies as well. To that struggle Biology ascribes the progressive evolution of the animal world. History takes the same line of argument; and political economists, in their naive ignorance, trace all progress of modern industry and machinery to the 'wonderful' effects of the same principle. The very religion of the pulpit is a religion of individualism, slightly mitigated by more or less charitable relations to one's neighbors, chiefly on Sundays. 'Practical' men and theorists, men of science and religion, preachers, lawyers and politicians, all agree upon one thing—that individualism may be more or less softened in its harshest effects by charity, but that it is the only secure basis for the maintenance of society and its ulterior progress.[1]

For those to whom the word "anarchy" means only cold-blooded terrorism, it may come as a surprise or even a shock that Peter Kropotkin, once a Prince of Russia and page to the tsar, is one of the internationally recognized scholars whose works are foundational to modern anarchist philosophy. Anarchy—as the restriction of large-scale, centralized government—does *not* imply chaos, as is usually charged; rather, it demands cultural coherence and social stability based on individual and cooperative self-regulation. The careless use of the term anarchy (if careless it is) by the politicians and media of industrial civilization has wrapped the beautiful Greek word *anarchia* in the stinking rags of ideological ranting; the real reason for debasing the word—and, implicitly, its partisans—is that anarchy demands systematic decentralization of power, a proposal against which all large-scale and centralized systems stand opposed.

As Kropotkin shows, the rise of the state, forcing into existence "loose aggregations of individuals, connected by no particular bonds," brings about the atomized individual in the modern sense. It is out of atomized desperation that terrorism, whatever its immediate aims or ideology, is born; and it is a semantic (and political) travesty to equate anarchy with terrorism, as has happened in contemporary usage. This process of language degradation parallels the destruction of organic cultural life. Both the all-powerful state and the "autonomous" individual emerge from the destruction of traditional community. If society (in the sense that Kropotkin uses the word) disappears, then both community and society drop out of the sequence individual-family-community-society-state, and we are left with individual-family-state. More accurately, however, we have the following series of terms: individual-family-bureaucracy-state. That is, the institutional connections of bureaucratic relationships (personnel) are advanced civilization's feeble substitute for both community and society.

In his *Eros and Civilization*, Herbert Marcuse has said that "The independent family enterprise and, subsequently, the independent personal enterprise cease to be the units of the social system; they are being absorbed into large-scale impersonal groupings and associations."[2] This description is most certainly true; but Kropotkin was pointing to a deeper social history than family or individual "enterprise," as we generally use and understand the word. Kropotkin's intention was to recall to us an organically functioning *community* grounded in local self-provisioning and only incidentally composed of small producers for the commodity market. This is not to say that we need a marketless society like the so-called "dark ages" in Western Europe after the fall of Rome. But the evolution of culture has been embedded, historically speaking, in production for use. To undermine this evolution by the enforced imposition of commercial organization leads to cultural sterility and, eventually, to organizational breakdown. The family farm, for instance, under the impact of agribusiness compulsions, tends to resemble more the business corporation than the self-provisioning homestead. Thus we read of farm families who, in the throes of foreclosure, don't have enough to eat. This is outrageous.

In the seventeenth century, the English political philosopher Thomas Hobbes provided the secular political anthropology, derived from aristocratic imperialism wedded to the legalistic orthodoxy of Judeo-Christian original sin, by which the state of nature was depicted as each against all, atomized terror that the political state drew forcefully out of chaos by the power of its orderly will. Civilization, with its royalty and aristocracy, has always held itself to be superior; its sword was not for crass exploitation but for the required maintenance of proper order; its affluence and luxury were proof of its human superiority and divine approval; it was—*noblesse oblige*—mandated to carry the cultural burden of elegant superiority; its very pride was humble obedience to God. The problem with this view is that its anthropology is not only false but simultaneously a thorough rationalization for the higher virtue of civilized exploitation. Have no illusion: civilization has always existed by, and sustained its existence through, expropriation. What did the word "tax" mean to millions of peasants down through the ages? To "democratize" civilization by bringing us all into the standardized largesse of global imperialism is to forget what every peasant always knew: at the bottom of the social heap somebody is raising food and getting it taken away. To be revolutionary now is to first dissolve the civilized imperialism we still have reigning in our minds.

So long as people can be made to believe that support for the personal and family enterprise is at the heart of the system's objectives, while the system in its actual functioning thwarts individual freedom and makes impracticable family subsistence, then people will uphold the system even as it crushes the very ideals it claims to promote and serve. This is a cruel paradox; and to see how deeply we have come, as a people, to support and uphold the system while it erodes what little remains of common culture is a sobering experience. It tells us of the power of myth and ideology in our lives and of our need to belong to an overarching social entity, even if that entity—the industrial system—steadily devours the common culture that, in the long run, makes coherence possible.

As if to protect ourselves from the internal crisis associated with recognizing the system's true objectives, many of us withdraw into our family lives, replete with institutional dependencies and mythological rationalizations, and refuse to engage in any social involvement that carries tension or controversy. Only associations that soothe anxiety are sustained. Outside the church or the hobby group, the standard form these associations take is the job; hence the tendency to fit in with the chronically repressed unhappiness, the friendly, amiable, tranquilized hypertension of personnel. Those who are terrified of community and mutual aid are not necessarily afraid to sell their lives piecemeal to bureaucracy, just as soldiers will consent to kill or be killed in battle rather than refuse to obey, even if the overarching cause for which they are fighting is predatory and corrupt.

One hundred and fifty years after the publication of *Walden*, we can still say with Henry Thoreau that most people continue to live lives of quiet desperation. Only in our time the advances in alienation—the greater distances from both nature and community—bring quiet desperation to the brink of that social paralysis that frightened George Orwell into writing his *1984*. Mass political passiveness, in other words, clouded and distracted by assurances from the system, its religion, and its science, leads us to the door of technocratic fascism. Whether we go through that door—as surely to our own doom as to the doom of Earth—depends on our ability and our willingness to transcend our ignorance and timidity, assume our cultural responsibilities, and accept our rightful place in the evolution of spiritual ecology.

ENDNOTES

1. Kropotokin, *Mutual*, 224–28.
2. Marcuse, *Eros*, 96.

6

Putting Our Minds Together

IN AN ARTICLE PRINTED several years ago in the Mohawk Nation newspaper *Akwesasne Notes*, a writer by the name of Sotsisowah (also known as John Mohawk) pointed to the Industrial Revolution as *the* historic convulsion that shattered traditional social groupings and cultural patterns. Sotsisowah says the Industrial Revolution "truly revolutionized much of the basic structure of society." The family, he continues, "was the major unit of production, at least in most rural and semi-rural societies," but once the industrial system gained power, the "function of production gradually left the home." The loss of this function, along with "educational, judicial, and religious/psychological functions," spelled a slow but sure death for traditional communities.[1]

Yet there seems to be a contradiction between Sotsisowah's analysis and his remedy. He quotes Thomas Porter ("a spokesman for the traditional people at Akwesasne") and the two paragraphs attributed to Mr. Porter are in italics to emphasize their importance. The main idea is also expressed in bold type beneath a drawing of an Indian family (father, mother, and son) looking at an infant held by the mother. The caption reads, "And the two of them are supposed to climb into a harness, just like a team of horses, and pull together, all the way through life."[2]

Sotsisowah insists the "Western nuclear family is far from the norm. Of the hundreds of cultures in the world, the West may be the only culture in which the nuclear family is considered the norm. The Western family is unique. It is also not functioning very well." What's puzzling here is that the "team of horses" image, that Sotsisowah implicitly advocates, is precisely the nuclear family; and *that*, he says, is "not functioning very well."[3] That is, he shows the nuclear family to be the product of industrialization (of which he is highly critical, and rightly so) but then offers up the nuclear family as the best corrective. How Sotsisowah can hold to both ideas at once is a little bewildering.

The nuclear family is the social husk left over when industry, in the words of Karl Polanyi, smashed "social structures in order to extract the element of labor from them."[1] The nuclear family provides the essential functions that the industrial economy requires: participation in industrial production, limited reproduction, and maximum consumption of mass-produced commodities. The nuclear family is the residue left over from the destruction of preindustrial society; with its wage work on the one hand, and its unpaid housework on the other, the nuclear family is the unstable compromise between nonindustrial community and an utterly privatized space-age world where nothing but the polluted air is free and where all other needs and desires can be accommodated in the market-place—provided one's computer records are in proper order and so long as one's "personalized" computer card or chip is not mashed, mutilated, mangled, or otherwise defaced.

The nuclear family in advanced industrial society, in other words, represents a pattern of life that is utterly dependent on the ongoing "prosperity" of the industrial economy. Carried to its logical conclusion, this trend toward total psychological investment in the status quo would seek to eliminate any vestige of community; indeed, community would be, by its very existence, a form of subversion. (Or, alternatively, the word "community" is used corruptingly, as in "defense community," "intelligence community," or "banking community.") "Family values," as a penetrating political slogan, seems to glorify the family on the surface, but its buried inner dynamic activates anxiety against contamination by anything openly public, thus sending hidden danger signals in regard to engagement with multicultural democratic life. Fearful religious righteousness oozes in and around the edges of such rhetoric. Perhaps we might paraphrase Kropotkin by saying that *families* now represent loose aggregations connected by no particular bonds, except in so far as they may prefer Pepsi over Coke, football over baseball, the Air Force over the Marines, game shows over soap operas, or James Dobson over Pat Robertson. It may be that Orwell's *1984* is a pushing to imaginative limits a trend that could never be attained in political reality. Perhaps; but we would do well to meditate on the *actual* political climate created under Stalin and Hitler.

---

1. In the case of indigenous American cultures, however, the smashing was not to extract labor but to expropriate land.

Our cheerful optimism in regard to technological freedom may be neither as justified nor as deserved as we would like to believe.

In the June 1997 *Reader's Digest* there was an article, by Michael Kinsley, called "Orwell Got It Wrong." Kinsley worked as publisher of an on-line magazine called *Slate*. He opened his article by telling the reader that George Orwell's prediction, in *1984*, that "technology would become the tool of totalitarian dictatorship" was "exactly wrong." Instead, Kinsley says, the "high-tech devices that have invaded our lives—home computers, fax machines, VCRs and now the Internet—have *expanded* human freedom." (Emphasis in the original.) He went on to say that "today's technology revolution is also a revolutionary advance for human liberty."[4]

I quote all this because, as electronic surveillance becomes normative and as its capacity for even greater invasiveness enlarges, electronic communication via computer and cell phone has become our virtual new religion. Impulse communication becomes "freedom." Somehow the *capacity* for real hard-nosed totalitarian dictatorship (as opposed to the mere imperialist market dictatorship we live under at present) seems beyond the mental ken of technological enthusiasts like Kinsley. (Jerry Mander, in a chapter called "Seven Negative Points about Computers," insists that "In terms of everyday life, the greatest danger of computers may be the level of surveillance they make possible. Computers have enabled the major institutions of our society—corporations, government agencies, the police, the military—to keep records well beyond what was previously possible."[5]) When real crises hit the United States, crises that truly disrupt the familiar routines of our everyday lives, crises that disable the market economy, Michael Kinsley's sanguine assertions will be put to the historic test. And we shall see whether he or George Orwell has the more prophetic soul. More importantly, we will only then fully realize what a staggering price there is to pay for removing the real "safety net" that has heretofore always existed to cushion the follies and tragedies of civilized disaster. I mean small-farm culture, homesteads, the commons, and farming-as-a-way-of-life in a countryside that knows the meaning and practice of sufficiency. As a wiser Marie Antoinette might say of our current crop of fervent techno-wonks, "Let them eat communication."

Because, as a nation, we do not know what it means to suffer, our self-righteous innocence and smug sense of superiority are seen as cruel imperiousness by those who do know the meaning of suffering and the reality of pain. In the parlance of the Christian faith, only those who know the reality of grief and the meaning of shame can be saved; those who are convinced of their righteous salvation are not yet ready for conversion to earthy humility. And to those who assert that such thoughts only represent unenlightened fear mongering or a low self-esteem affection for dumbing down, it must be said that a civilization capable of developing a policy of Mutually Assured Destruction (whose acronym itself is testimony to the theater of the spiritually insane, the absolute opposite of loving, embracing, or forgiving your enemy) is also capable of squeezing the cultural vitality and communal coherence out of society at large, and of substituting narcotic distraction for real fulfillment. What is urgently needed now is not so much nervous fretting over divorce statistics as steadfast commitment to building stronger communal ties and associations. The family quite obviously plays a strong and decisive part in communal coherence; but there is only harness enough for two in a team. It is precisely the *limitations* of the nuclear family that need to be transcended.

In the present context, however, the nuclear family serves the purposes of industrial society; its desperately busy, privatized quietism provides the human oil on which the industrial Leviathan can run unobstructed. Bereft of community, the family becomes an agency for consuming not only the goods and services of industrial production but commercial culture itself. That is, the very culture of the people is a commodity up for sale, and its price is chalked up to the GNP. The commodity orientation of the industrial family is itself a key obstacle to a more whole and more deeply connected way of life. In terms of the ideology of GNP, any call for reduced consumption is suspect if not subversive, for reduced consumption threatens the comfort level of our distractions, and the prospect of being thrown back on our own resources creates instant anxiety. Yet comfortable, private families tend to withdraw from cooperative associations and stay within the parameters of the system, thus maintaining the atomizing process.

Near the end of his article in *Akwesasne Notes*, Sotsisowah says that people are now "specialized to the needs of modern technological industrial society" and are no longer "specialized to the needs of the human species specifically."[6] Even here the use of terms is revealing: we are not so much in need of people who are "specialized" to the needs of "the human species" as we are in need of a society in which people are *generalized* in skills and aptitudes for the sake of their communities and the wholeness of their persons.

Sotsisowah also says that the "survival of the species depends on the survival of the young."[7] Yet the survival of the young is rendered greatly uncertain by virtue of our cultural disorientation; the industrial system, so long as it commands our allegiance, keeps us from knowing how lost we really are. There is no such thing, finally, as sheer physical survival without a cultural and communal context that provides meaningful direction for our lives; this is, despite their sincerity, the serious misapprehension of "survivalists"—promoters of romantic and heroic individualism.

The survival of the species depends as much on cultural continuity as it does on physical reproduction. Without the former, the later becomes meaningless. This fact may help to explain the phenomenon of declining birth rates in advanced civilization: when the cultural fabric has been rendered threadbare, child-rearing itself becomes a form of institutional anguish, generating whole departments in the education and social service "industries."

II

The intuitive turning toward the young has been the compelling force that has led to the creation of "alternative" schools. These schools are generally organized in such a way so as to promote child-centered education, emphasizing personal relationships as much as skill learning. An attempt is thereby made to promote or develop community from within the classroom or school context.

"Let us put our minds together and see what kind of life we can make for our children"—so said the legendary Sioux Indian Sitting Bull, as quoted by Sotsisowah.[8] These words express simply and powerfully an important adult responsibility. With some sensitivity to the Indian experience, we should be able to appreciate the insight that an adequate educational process must derive from community, and that the attempt to squeeze community out of education is a rather misguided proposition. To complain that our schools are failing to educate our children adequately is to miss the essential point. Such complaints contain an unexamined assumption: that schools, rightly honed, are capable of offering education in depth. But human beings who are processed through a bureaucratic machine will become educated in spite of the machine, not because of it. The educational apparatus can program our youth, but it cannot—indeed it will not—set them free. Little wonder, then, that the computer is now proclaimed the educational messiah; such programming is the analogue for what the school system accomplishes with our children.

As long as people continue to believe that a more liveable world is contingent on adjusting the curricula of the school system to industrial expectations, whether flatly for "jobs" or more euphorically for "the frontiers of high technology," the net outcome of such efforts will only be more regimentation—even if the regimentation becomes increasingly voluntary and self-imposed. So long as adults deflect onto their children's lives the "solutions" for a more liveable world, the world will only become more unliveable. Only when adults demand and struggle for a depth of self-regulated cooperative structure *in their own lives* will the educational machine, and the industrial apparatus whose function it serves, begin to receive serious challenge. Adults who haven't themselves the courage to stand in opposition to corporate power and the bureaucratic system, who won't risk voluntary "poverty" or an alternative life style, but who secretly (if not unconsciously) hope to enable their children to be brave enough to live self-directed, purposeful lives through the medium of "rich environments" or "gifted and talented" programs, are engaged in a nearly hopeless fantasy. In a world where fulfillment is endlessly deferred or sublimated, only a stand for personal integrity and communal coherence has any sustained relevance. To push "the future" off onto our children is to give the future into the hands of bureaucratic technicians. The motive for so doing may be based on a desperate hope, but it is enacted through an unexamined fear. Courage cannot be taught, only demonstrated.

In his article, Sotsisowah misses the deeper point, namely that cultural coherence is dependent on the integrity of community. The coherence of preindustrial community was largely unconscious and taken for granted. But once community is shattered, its reconstruction depends on consciousness, clarity, and commitment. In the Introduction to Schumacher's *Small is Beautiful*, Theodore Roszak provides implicit historical support for this insight:

> Schumacher's work belongs to that subterranean tradition of organic and decentralist economics whose major spokesmen including Prince Kropotkin, Gustav Landauer, Tolstoy, William Morris, Gandhi, Lewis Mumford, and most recently, Alex Comfort, Paul Goodman, and Murray Bookchin. It is the tradition we might call anarchism, if we mean by that much abused word a libertarian political economy that distinguishes itself from orthodox social-

ism and capitalism by insisting that the *scale* of organization must be treated as an independent and primary problem. The tradition, while closely affiliated with socialist values, nonetheless prefers mixed to 'pure' economic systems. It is therefore hospitable to many forms of free enterprise and private ownership, provided always that the size of the private enterprise is not so large as to divorce ownership from personal involvement, which is, of course, now the rule in most of the world's administered capitalisms. Bigness is the nemesis of anarchism, whether the bigness is that of public or private bureaucracies, because from bigness comes impersonality, insensitivity, and a lust to concentrate abstract power. Hence, Schumacher's title, *Small is Beautiful*. He might just as well have said 'small is free, efficient, creative, enjoyable, enduring'—for such is the anarchist faith.

Reaching backward, this tradition embraces communal, handicraft, tribal, guild, and village life-styles as old as the neolithic cultures. In that sense, it is not an ideology at all, but a wisdom gathered from historical experience.[9]

Smallness of scale was the dominant social pattern in preindustrial life; to reconstruct smallness in the present period depends on personal consciousness and political will. At some level, each of us has to *choose* reform by retrogression.[II] The wisdom of this perspective must now assume political expression, although it is obviously no easy task to resurrect folk culture, now largely broken, to a new and fuller expression free from civilized institutions and therefore capable of digesting the predatory features within civilization: transforming civilization's utopian swords into eutopian plowshares, its nuclear spears into solar pruning hooks.

In the past, a community could survive the loss or breakup of a family—the overall way of life would go on—but a family could not survive the loss of community without its essential way of life being wrecked. To idealize the family, as Sotsisowah does, is to fall into the trap of individualism, of "family values." This gets us nowhere. The family would not have to be idealized (and, by being idealized, its problems glossed over) if it were solidly embedded in living community. The compulsive cloistering now so common in nuclear families is evidence of a lack, and not by any means proof of an abundance, of family joy. Cloistering only confirms

---

II. For a refreshing glimpse into the governing process of the Iroquois Confederacy and the Hopi, see "The Gift of Democracy," Chapter 13 in Jerry Mander's *In the Absence of the Sacred*.

the old saw about fanaticism: when the aim is lost, the effort becomes maximized. Much of the energy pumped into alternative schools would be much better spent, even in terms of education for children, if adults worked at and worked out their own community.

We will never get a liveable world by deferring our needs onto our children. Despite the best of intentions, what we are really teaching them is pampered deferment and timid postponement. Despite all our political rhetoric about bravery and freedom, we are unwilling to face into conflict and afraid of facing up to intimacy. We are fenced in by fear.

ENDNOTES

1. Sotsisowah, "Future," 5.
2. Sotsisowah, "Future," 6.
3. Sotsisowah, "Future," 6.
4. Kinsley, "Orwell," 131.
5. Mander, *In the Absence*, 63.
6. Sotsisowah, "Future," 7.
7. Sotsisowah, "Future," 6.
8. Sotsisowah, "Future," 4.
9. Roszak, "Introduction," 3–4.

7

The Educator's Engine

IN *NEWS FROM NOWHERE*, a late nineteenth-century novel by William Morris, there is an interesting dialogue I will shortly quote. But first a little background. *News from Nowhere* is eutopian. Morris wrote *News from Nowhere* in response to a different sort of utopian novel, *Looking Backward*, written by Edward Bellamy. Bellamy's novel depicted a regimented consumer paradise, and Morris was deeply irritated by what he read. Morris's vision of a desirable society was full of spontaneity and physical delight. Instead of passive consumerism, Morris wanted creative initiative; instead of writing about a sunless anthill metropolis, Morris cast his characters in the countryside. And, as the reader shall see, Morris's tale is filled with implications for education.[1]

As *News from Nowhere* opens, the author, as first person character, has fallen asleep in his bed (in Hammersmith, a London suburb, in 1890) and awakens in the twenty-first century. He finds himself in the same location, except that everything, including the building in which he went to sleep, is new and different. After an episode of discovery and introduction—an experience as puzzling to the author as to his amused hosts and hostesses—he gets a guide by the name of Dick Hammond; and as they are traveling the following day (by horse and carriage!), our author sees many strange and transformed sights. A group of children, tenting in meadows near the edge of a forest, sets the following scene:

> Romantic as the Kensington wood was, however, it was not lonely.
> We came on many groups both coming and going, or wandering
> in the edges of the wood. Among these were many children from
> six to eight years old up to sixteen or seventeen. They seemed to
> me to be especially fine specimens of their race, and were clearly

1. For a fuller explication of the eutopian/utopian dynamic, see my *Green Politics Is Eutopian*, also published by Wipf and Stock.

enjoying themselves to the utmost; some of them were hanging about little tents pitched on the greensward, and by some of these fires were burning, with pots hanging over them gipsy fashion. Dick explained to me that there were scattered houses in the forest, and indeed we caught a glimpse of one or two. He said they were mostly quite small, such as used to be called cottages when there were slaves in the land, but they were pleasant enough and fitting for the wood.

'They must be pretty well stocked with children,' said I, pointing to the many youngsters about the way.

'O,' said he, 'these children do not all come from the near houses, the woodland houses, but from the countryside generally. They often make up parties, and come to play in the woods for weeks together in the summertime, living in tents, as you see. We rather encourage them to it; they learn to do things for themselves, and get to notice the wild creatures; and, you see, the less they stew inside houses the better for them. Indeed, I must tell you that many grown people will go live in the forests in the summer; though they for the most part go to the bigger ones, like Windsor, or the Forest of Dean, or the northern wastes. Apart from the other pleasures of it, it gives them a little rough work, which I am sorry to say is getting somewhat scarce for these last fifty years.'

He broke off, and then said, 'I tell you all this, because I see if I talk I must be answering questions, which you are thinking, even if you are not speaking them out; but my kinsman will tell you more about it.'

I saw that I was likely to get out of my depth again, and so merely for the sake of tiding over an awkwardness and to say something, I said—

'Well, the youngsters here will be all the fresher for school when the summer gets over and they have to go back again.'

'School?' he said; 'yes, what do you mean by that word? I don't see how it can have anything to do with children. We talk, indeed, of a school of herring, and a school of painting, and in the former sense we might talk of a school of children—but otherwise,' said he, laughing, 'I must own myself beaten.'

Hang it! thought I, I can't open my mouth without digging up some new complexity. I wouldn't try to set my friend right in his etymology; and I thought I had best say nothing about the boy-farms which I had been used to call schools, as I saw pretty clearly that they had disappeared; so I said after a little fumbling, 'I was using the word in the sense of a system of education.'

'Education?' said he, meditatively, 'I know enough Latin to know that word must come from *educere*, to lead out; and I have heard it used; but I have never met anyone who could give me a clear explanation of what it means.'

You may imagine how my new friends fell in my esteems when I heard this frank avowal; and I said, rather contemptuously, 'Well, education means a system of teaching young people.'

'Why not old people also?' said he with a twinkle in his eye. 'But,' he went on, 'I can assure you our children learn, whether they go through a "system of teaching" or not. Why you will not find one of these children about here, boy or girl, who cannot swim; and every one of them has been used to tumbling about the little forest ponies—there's one of them now! They all of them know how to cook; the bigger lads can mow; many can thatch and do odd jobs carpentering; or they know how to keep shop. I can tell you they know plenty of things.'

'Yes, but their mental education, the teaching of their minds,' said I, kindly translating my phrase.

'Guest,' said he, 'perhaps you have not learned to do these things I have been speaking about; and if that's the case, don't you run away with the idea that it doesn't take some skill to do them, and doesn't give plenty of work for one's mind; you would change your opinion if you saw a Dorsetshire lad thatching, for instance. But, however, I understand you to be speaking of book-learning; and as to that, it is a simple affair. Most children, seeing books lying about, manage to read by the time they are four years old; though I am told it has not always been so. As to writing, we do not encourage them to scrawl too early (though scrawl a little they will), because it gets them into a habit of ugly writing; and what's the use of a lot of ugly writing being done, when rough printing can be done so easily. You understand that handsome writing we like, and many people will write their books out when they make them, or get them written; I mean books of which only a few copies are needed—poems, and such like, you know.'[1]

There are critics who accuse Morris of being a perfectly hopeless romantic, a man whose alleged nostalgic idealization of preindustrial life was simply projected into the future through the medium of art. It is true that Morris, like his mentor John Ruskin, was heavily influenced by medieval folk culture and craftsmanship. But to honestly honor excellence and freedom in quality craft is certainly something different, and something greater, than nostalgia. Morris gets labeled a romantic because he sees too clearly for our comfort. He is unwilling to participate in the posture of sardonic consumerism.

If, as I have asserted, our present educational system is shaped to the structural needs of the industrial economy rather than to the real needs of vigorous children, then it follows that the platitudinous endorsement of compulsory schooling by educational "experts" may be more a matter of ideological conformity than wise educational counsel. Just how deeply this conformity runs in the educational establishment is given unsettling expression by the late Paul Goodman in his *Compulsory Miseducation*:

> It is uncanny. When, at a meeting, I offer that perhaps we already have too much formal schooling and that, under present conditions, the more we get the less education we will get, the others look at me oddly and proceed to discuss how to get more money for schools and how to upgrade the schools. I realize suddenly that I am confronting a mass superstition.[2]

Goodman may be right (I believe he is) in calling "superstitious" the attitudes he confronts; but we would be intellectually negligent if we failed to recognize the extent to which this educational superstition is also a highly charged and intensely rationalized appendage to industrial ideology. Industrial civilization asserts its values simultaneously as normative and ideal; it crushes that which opposes or contradicts its standardizing expression. This process is "inevitable" only to the extent that supposedly freely held and rationally weighed values are, in intellectual reality, in the service of "mass superstition." The less we know how to do for ourselves, the more we are told of our need for extended education. The more we become convinced of our "backwardness" and ashamed of our origins, the more vulnerable we are to predatory "enlightenment."

Rising population in an urban-industrial world where true human labor is less and less in demand, where automated machines make tools obsolescent and where culture is itself a commodity to sell and buy, creates a vertical extension and horizontal expansion of school systems; this growth runs parallel to the extension and expansion of industrial technology. The school system is a structural reflection of the industrial economy. As young people are needed less and less for any productive purpose (and who, by all means, are not to be "condemned" to the "obsolescent" skills of their ancestors), their inherent need for real and useful competencies goes undeveloped. The school system, making the air blue with all manner of empty rhetoric, sloganeering, and concept-divination, picks up the social slack with prolonged babysitting. (What else can it be called when an eighteen-year-old person, old enough to fight and die in the military, must ask permission to go to the toilet?)

The techno-industrial need for increasingly specialized specialists has been fuel for the educator's engine: the greater the number of specialists required, and the more precise their specializations become, the more the various educational systems can and do expand. The higher the pyramid rises the broader the base must be; millions of young people are compelled to "consume" twelve to twenty years of *continual* organized schooling. That most of this schooling has little to do with meaningful education in a naturally interesting world, and largely to do with controlling the time of youth for the sake of the industrial treadmill, as an "investment" in technological progress, is obvious to any honest observer. Except for the rare few who are able to appropriate selected features of the school curriculum to their own advantage, the vast majority of those who graduate from high school or even college have neither the competence nor the organic self-confidence that comes with the mastery of real skills for living. In its merciless scanning of youth for those inclined to exacting industrial specializations (chemistry, math, physics, electronics, computers), the industrial school system keeps all our children from self-directed experience and a less problematic maturation. People who have no useful competencies are destined either to chronic anxiety or stylistic posturing. Such persons may be extraordinarily malleable in rapidly changing "roles" or compulsive "retraining," but quiet self-direction becomes an elite privilege rather than a common right. People become personnel.

Rapid invention, industrial mass production, and technological specialization have virtually eliminated most traditional occupations by undermining both the methods and the cultural embeddedness of those occupations. In the modern industrial world, one is virtually forced to become a specialist of some sort; those who refuse, consciously or unconsciously, are doomed to a marginal existence—except, once again, for those rare few whose strength of will and clarity of moral purpose enables them to live more self-directed lives, even as such self-direction often includes a willingness to embrace so-called "voluntary poverty." Our task is to democratize self-regulation and so rebuild a common and earthy culture.

When the culture of community is abolished and made impracticable, the training of the young falls to standardized, compulsory institutions. Because the cultural fabric has been rendered so threadbare, the rearing of children itself becomes problematic. Parents intuitively realize that there is little in their commodity-intensive lives that has inheritable cultural value. Social workers, therapists, and counselors become the privileged class who understand "roles" and help people "adjust" and "adapt." These are the new technicians of civilized behavior.

Without a close network of communal subsistence, the nuclear family has degenerated into a mere consuming household whose members only wish to be entertained simultaneously on different channels. In such a domestic context, where virtually all cultural stimuli come through commercial, bureaucratic, or electronic conduits, parents rely on the school system to take their children off their hands, to unburden themselves from the culturally unbearable weight of their own progeny. This is true, by and large, of parents who don't work as much as those who do. (Of course "work" is what one does *away* from home, where the presence of family members would be an embarrassment, a nuisance, or completely prohibited by corporate rules or insurance restrictions. And the contrived enthusiasms for "family projects," cheerful to the bitter end, are psychological tragedies.)

There is something exceptionally peculiar here. The school is not so much an intrusion into the life of the household as home life is an utterly unproductive, idle interlude in the repetitive rigors of the classroom. A professor of education from the University of Massachusetts who spoke at a conference on education in St. Louis, Missouri, was quoted in the *St. Louis Post-Dispatch* as saying the home is an "untapped educational resource," and ways must be found to "utilize" this resource and "bring it into the system." This kind of thinking represents the leading edge of industrial-educational expertise.

Bureaucratic organization and organic community are not only structures of a *different* nature, they are inherently in *opposition*, especially in so far as bureaucracy manifests the civilized drive to overpower and suppress folk life, to smother it in compulsory "caring." The increasing reliance on standardized systems is a clear and unmistakable indicator of the erosion of folk community and common culture. Joseph R. Strayer, in his *Western Europe in the Middle Ages*, says that just as the Roman Empire "succeeded in creating a real community of interest and feeling in the Mediterranean basin, it began to decay. This decay is one of the great puzzles of history, and no one has ever been able to explain it in a completely satisfactory way."[3] Strayer then goes on to describe in some detail how the imperial state had destroyed the organic tissue of society. He even says that "We can say with absolute truth that the Roman Empire fell because the great majority of its inhabitants made no effort to preserve it. They were not actively hostile to the Empire; they were merely indifferent."[4]

The decay of Empire or an imposed organizational system seems to me no great puzzle at all. Such organization becomes a rational suppression of organic culture, and to those on whom its demands are placed its perpetuation is at best a matter of indifference, especially when the imposition is structurally predatory, as has overwhelmingly been the case with civilization to date. Writing of Roman collapse, Joseph A. Tainter in *The Collapse of Complex Societies* said:

> The burden and costliness of the Empire not only increased over time, but the benefits it afforded its members declined. As crops were confiscated for taxation and peasants' children sold into slavery, lands were increasingly ravaged by barbarians who could not be halted with the Empire's resources. The advantage of empire declined so precipitously that many peasants were apathetic about the dissolution of Roman rule, while some actively joined the invaders.[5]

This is not to say that there is no place for organization in the conduct of life, whether educational or governmental. The issue is, in global terms, how to achieve a transcultural unification without extirpating the embedded diversity on which the unification depends, or, alternatively, how to achieve and sustain the humane eutopian universality civilization says it desires but which its vicious utopian methods obstruct and prevent. Bureaucratic organization devours common culture. To bring the "untapped resource" into the system is to destroy the culture that stands in the way. To have a society that is so totally hollowed out is to be left with a standardized totalitarianism—a residual fascism of the corporate state.

II

There is among alternative school educators a conviction (not universally accepted) that alternative or progressive education is radically different than the kind of education found in the compulsory system. It is certainly a conviction open to challenge and subject to debate. Let us look briefly at one of the modern pillars of alternative-progressive education: John Dewey's *The School and Society*. In this early work by Dewey, there are a great many insights. I will quote a few:

> Unless culture be a superficial polish, a veneering of mahogany over common wood, it surely is this—the growth of the imagination in flexibility, in scope, and in sympathy, till the life which the individual lives is informed with the life of nature and society . . . In the first place, the ideal home has to be enlarged. The child must be brought into contact with more grown people and with more children in order that there may be the freest and richest social life . . . The unity of all sciences is found in geography . . . The radical reason that the present school cannot organize itself as a natural social unit is because just this element of common and productive activity is absent . . . Upon the ethical side, the tragic weakness of the present school is that it endeavors to prepare future members of the social order in a medium in which the conditions of the social spirit are eminently wanting.[6]

There is a great deal of substance here. We can easily see that Dewey links culture with nature and with earthy society; he identifies the "tragic weakness" of the school system with a general absence of the "social spirit." He is advocating the creation of a pattern of schooling that would both promote and enable a sustained interest in the natural world; he approves of passion, enthusiasm, and spontaneity; he wants the school to be fully integrated into the community, both human and natural. All this is exceptionally sound and exciting.

But this isn't all. Let us look at another passage from *The School and Society*—a passage that nullifies, absolutely and unconditionally, all of the above insights and proposals:

> Until the instincts of construction and production are systematically laid hold of in the years of childhood and youth, until they are trained in social directions, enriched by historical interpretation, controlled and illuminated by scientific methods, we certainly are in no position even to locate the source of our economic evils, much less to deal with them effectively.[7]

It's hard to believe the same person could express both clusters of ideas—in the same book! It is difficult to understand how Dewey, the eminent philosopher, could be so philosophically slippery and so obeisant to "scientific" ideology. What a conception of childhood he presents: a blob of formless protoplasmic clay on an educational assembly line being punched and pressed into civilized shape by various scientific treatments—trained, enriched, controlled, and illuminated! Until the techniques of education and schooling are perfectly refined, he is saying, until science has complete control of educational methodology, we will have no way of knowing anything about our "economic evils," nor will we know how to "deal with them effectively." The scientist, the technician, the specialist will tell us once they know.

But a scientific methodology that is permitted to "lay hold of" the instincts of childhood and youth and transform them into a shape compatible with the scientific-industrial agenda leads insensibly to technological fascism: the complete administration of culture. No matter how many and how frequently the sops thrown to the "social spirit," they are in the end for nothing—unless people rise (preferably in a Gandhian way) to rebellion and revolt. It is the instincts that must lay hold of and control organization, not the other way around: and by "instincts" I mean not only esthetic intuition but the accrued traditional wisdom of the human race, certainly including the deepest spiritual ethics of the world's great religions. If there are things in civilization worth preserving—and I believe there are—it will be, in the end, traditional folk wisdom, esthetic intuition, artistic sensibility, and the deepest spiritual ethics that will provide the foundation of our collective judgment. Science may well be called upon and asked for its informed opinion, but it is a long way from joining the company of the venerable.

I am not a close enough student of Dewey to know how he resolved (or failed to resolve) this intellectual schizophrenia in his later life. There was, to be sure, a naive enthusiasm for science at the end of the nineteenth century, an enthusiasm that many intellectuals later reconsidered, especially after the appalling butchery of the First World War. But the ongoing political impact of such powerful but shallow optimism has bequeathed to us the modern school system, replete with all its unchallenged scientific superstitions, while the more practical and social aspects of progressive education have been systematically jettisoned as mere nostalgic backwardness, as trifling and piddling, as distractingly unfit for a *real* industrial society.

In short, the choice for scientific supremacy, although it may once have seemed the promise of the future, can now be recognized as collusion with intellectual and political cowardice, an obsequious rite of passage to the inner sanctum of the technocratic system. Insofar as this is the ideology of liberal scientific capitalism, it is where liberalism wets its pants.

ENDNOTES

1. Morris, *News*, 30–32.
2. Goodman, *Compulsory*, 10.
3. Strayer, *Western*, 14.
4. Strayer, *Western*, 20.
5. Tainter, *Collapse*, 188.
6. Dewey, *School*, 56, 36–37, 16, 11–12, 12.
7. Dewey, *School*, 22.

8

Technical Autonomy

L ET'S COMPARE THE TENDENCIES within the educational establishment to those operative in agriculture. The modern economic conception of agriculture is that of a food business, a commodity-producing industry. Since market efficiency and large turnover are hallmarks of any successful business, mass-producing techniques have been applied wholesale to agriculture. The old patterns of agrarian culture that stood in the way of this rationalization were blown away in the mechanical whirlwind.

Agrarian culture was nothing less than the slowly accrued life patterns of people who, for centuries, had built up stable and modestly productive methods for sustaining farm and village life. Whatever its limitations, shortcomings, and faults, this was a way of life that added up to a culture. It was capable of sustaining and perpetuating itself. It contained art, music, dance, clothing styles, and building methods. But in just a few generations, this culture has been systematically destroyed: where two hundred years ago, ninety-five percent of the counted population in the United States were engaged in farming, largely in a mode of self-provisioning, the situation at present is almost exactly inverted: less than two percent are now involved in farming, and only a fraction of those in dedicated householding.

The integrity and coherence of rural culture, a culture distinct in many ways from urban culture, have been thoroughly debased by the organization, technology, and economic demands radiating out of the commercial city. Radio, television, automobiles, consolidated school systems, agribusiness chemical and mechanical technologies, mass-produced houses and barns—all are aspects of what has been, within the past few generations, forcibly substituted for the community and craft of an older and more slowly paced rural life. Technology has allowed urban commercialism to penetrate the countryside as never before; the result is the disorientation and cultural enfeeblement of the countryside.

Cities may have traditionally been to the surrounding countryside what, in some limited respects, the brain is to the body. As long as their relative size and function were adequately differentiated, there could exist a healthy—if sometimes tense—symbiosis. But when the city passed beyond the point of being a balanced and balancing symbiont, it became an imperious parasite.[1] It is always in the interest of a parasite to see that its host remains healthy and functional: the parasite that consumes the host removes the basis, in the long run, for its own continuance. This is clearly the case for the industrial city; deterioration of prior civilizations shows collapse to be no idle probability.

Just as agriculture is not an industry, so too is community distinct from rational organization. Just as uncontrolled industrialization destroys the culture of the countryside, so too does the expansion of institutional organization serve to eradicate community. The more highly rationalized industry and organization become, the more quickly do they corrupt culture and devitalize community. Community, because it only comes, as Baker Brownell said, "like life, without machinery or artifice," must of necessity develop by the slow accretions of lived and living experience. Community develops in an organic matrix of time-honored tasks coupled with face-to-face interdependence in a knowable and liveable environment. Both school and farm could be, and ought to be, bastions of this kind of life. This is the spirit of Paul Goodman's educational ideal—to trust that our children are capable of growing up well with simple affections in a beautiful environment with freedom.

I. What we see with the Industrial Revolution is that civilization, as a governing construct, was able to relinquish its parasitic grip on the peasantry in favor of direct mechanical, chemical, and fossil fuel implements that eliminated the civilized need for a peasantry. The peasantry was therefore discarded. In its place were built direct instruments of technology and finance increasingly under the control of capital rationalization. With the rise of kingship, agriculture had fallen under continuous civilized expropriation. With the demise of hereditary aristocracy, the mythology and ideology of civilized superiority had so penetrated and saturated middle-class consciousness that civilized economics could now safely sleep in the same bed with democracy. "Democracy" and "development" became the salt and pepper, the yin and yang, of modernization, even if troublesome democracy (as in Iran in the early 1950s) had to be roughly stuffed back in a box so that a righteous, if also brutally repressive, program of development could carry the bright torch of civilized values without being tugged backward by mere popular will or human need.

The cruel irony is that we have the material means by which we might realize Goodman's ideal in rather short order; but, having the means, we are almost totally lacking in spirit and in faith. Politically, we are constantly being asked to have faith in the technological assault on the future; we are asked to have faith in "science"—not the science of humble wonder, but the science of technical mastery. This latter science is riddled with cynical belief, but it is devoid of faith. All it can do is create an increasingly dangerous technological monster.

II

Perhaps it seems both excessive and unfair to lay such blame for cultural malaise at the wheels of technology. Some will say it is not technology but the economic ideology behind technology that is responsible for the modern disorientation we have come to call "alienation." Technology in itself is not the problem, some will say; the only real issue is the use to which technology is put. In many ways I agree with this point of view; but the issues are complex, and there is no point in being hasty.

In the general mud-slinging international debate between Left and Right—a "debate" in which substantive issues of intellectual merit are rarely, if ever, discussed—there is an obscured point of concurrence. That point is the common conviction that industrial growth and technological expansion are signs of economic health and social security. In challenging technology, therefore, it is possible to address the point at which both capitalism and communism are fundamentally distorted, where both ideologies reveal their cultural and spiritual deficiencies, evade their human and ecological responsibilities, by hooking their shriveled and insecure inner spirits onto a dynamic outer mechanism whose sleek, compact perfectionism becomes the object of their admiration, adulation, and even of their worship. Modern socialism, which is a blend of public and private ownership, for the most part also adheres to the dominant industrial creed. Socialism, capitalism, and communism all share a materialistic bias whereby nonmodern and nonindustrial cultural forms must give way to the standardized or commercial.

Technology, as such, may not be the problem: a hand pump is also a form of technology, as is a splitting axe, a sewing machine, a typewriter, a butter churn, or a scythe. But small-scale technology—*appropriate* technology for cooperative and personal use—is a means, not an end. It enables us to live simply and well. By contrast, the present clamor for high technology does not mean that such technology, whatever it is, will really improve the quality of life or the social conditions of our culture. High technology has become an end in itself, an utterly irrational cargo cult, spewing high-tech trash in its wake. It is tied into the unrelenting pace of progress, to which we must give way as before fate or an act of God. The challenging of "technology" is thereby a way of getting at the underlying issues: whether people create culture in relation to Earth, or whether culture is the froth of the organizational machine as it speeds us toward an illusory technological perfection.

In the chapter that gives his book its title ("In the Absence of the Sacred"), Jerry Mander talks about "technical fixes that can accommodate continued industrial activity" (the "iron enrichment" of oceans, ozone "bullets" fired into the stratosphere, nanotechnology for redesigning molecular structure, etc.), but then goes on to "the ultimate technology nerds" who "speak about *downloading* their consciousness into computers." He also says:

> The greatest universities in this country—Stanford, MIT, Harvard, Berkeley, Princeton—provide these projects funding and housing and a platform to speak from. The United States military—particularly the Navy—backs many of these researchers with multi-million-dollar grants . . .
>
> All of these institutions can support these new modes of technological expression because the ideas are in every way consistent with the logic and the assumptions by which our society has operated for the past several centuries . . .
>
> The assumptions have been gaining strength for thousands of years, fed both by Judeo-Christian religious doctrines that have de-sanctified the earth and placed humans in domination over it; and by technologies that, by their apparent power, have led us to believe we are some kind of royalty over nature, exercising Divine will.[1]

Lewis Mumford traces all this back to the founding and foundation of civilization, to the institution of divine kingship. He insists that "scientific determinism not less that mechanical regimentation had their inception in the institution of divine kingship."[2] Can't we therefore see in the "downloading" of consciousness into computers nothing more than an updated form of the royal Egyptian mummy pretending toward an ersatz "immortality"?

III

Those who put the finger on technology are certainly not all to be found in the "counterculture." Solly Zuckerman, once the chief scientific adviser to the British Ministry of Defence and, later, chief scientific adviser to the British government, has some unkind things to say about technicians in his book *Nuclear Illusion and Reality*:

> It is my view, derived from many years of experience, that the basic reason for the irrationality of the whole process of the nuclear arms race is the fact that ideas for a new weapon system derive in the first place, not from the military, but from different groups of scientists and technologists who are concerned to replace or improve old weapons systems—for example, by miniaturizing components—or by reducing weight/yield ratios of nuclear warheads so that they can be carried further by a ballistic missile (that is to say, by packing greater explosive power into a smaller volume and weight). At base, the momentum of the arms race is undoubtedly fueled by the technicians in govermental laboratories and in the industries which produce the armaments ...
>
> In the nuclear world of today, military chiefs, who by convention are a country's official advisers on national security, as a rule merely serve as the channel through which the men in the laboratories transmit their views. For it is the man in the laboratory, not the soldier or sailor or airman, who at the start proposes that for this or that reason it would be useful to improve an old or devise a new nuclear warhead; and if a new warhead, then a new missile; and, given a new missile, a new system within which it has to fit. It is he, the technician, not the commander in the field, who starts the process of formulating the so-called military need. It is he who has succeeded over the years in equating, and so confusing, nuclear destructive power with military success. The men in the nuclear weapons laboratories on both sides have succeeded in creating a world with an irrational foundation, on which a new set of political realities has in turn had to be built. They have become the alchemists of our times, working in secret ways that cannot be divulged, casting spells which embrace us all. They may never have been in battle, they may never have experienced the devastation of war; but they know how to devise the means of destruction.

And the more destructive power there is, so, one must assume they
imagine, the greater the chance of military success.[3]

If, to military technicians, sheer destructive power correlates to
military success, then it appears equally true that the "standard of living,"
in the minds of civilized commodity technicians, depends on a continually expanding industrial productivity. These are two sides of the same
ideological coin. Instead of living in a world based on enriching culture
and personal fulfillment, in which our tax dollars are used to promote
ecological health and facilitate all manner of international understanding,
we exist instead in a condition of intellectual poverty, deluded by conceptions of sterile progress and drenched in illusions of how the "Christian
free world" must build more and more arms in order to overpower, or
defend itself from, potential enemies abroad. In our time, we are seeing
the truthfulness of the old Taoist aphorism—"Knowledge of Good and
Evil is a sickness of the mind"—revealed in its full subtlety. That which
comes in the garb of light is devoid of humanity.

The French sociologist Jacques Ellul, in *The Technological Society*,
says that "it seems impossible to speak of a technical humanism."[4] Ellul, a
Christian, goes on:

> [T]echnical autonomy is apparent in respect to morality and spiri-
> tual values. Technique tolerates no judgment from without and
> accepts no limitation . . . Morality judges moral problems; as far
> as technical problems are concerned, it has nothing to say. Only
> technical criteria are relevant. Technique, in sitting in judgment of
> itself, is clearly freed from this principal obstacle to human action.
> Thus, technique theoretically and systematically assures to itself
> that liberty which it has been able to win practically. Since it has
> put itself beyond good and evil, it need fear no limitation whatever.
> It was long claimed that technique was neutral. Today this is no
> longer a useful distinction. The power and autonomy of technique
> are so well secured that it, in its turn, has become the judge of what
> is moral, the creator of a new morality. Thus, it plays the role of
> creator of a new civilization as well . . . We no longer live in that
> primitive epoch in which things were good or bad in themselves.
> Technique in itself is neither, and can therefore do what it will. It
> is truly autonomous.[5]

Ellul goes on to elucidate how, for a long time, human conduct "belonged
to the realm of art":

Behavior based on flair, on intuitive as well as reasoned knowledge, and on personal relations; the spontaneous devising of means for influencing heart and mind; the wholehearted participation of man in his acts—all these are characteristic of art. Great leaders, great teachers, and agitators have all been artists. But art and artistry no longer suffice. We must find solutions to the problems raised by techniques, and only through technical means can we find them.[6]

To say that "art and artistry no longer suffice" is equivalent to saying that understandings and solutions no longer come out of accrued cultural wisdom but, rather, from the technical expertise that directs the political-economic machine. One need only consider the growing urgency regarding the disposal of radioactive wastes from nuclear weapons and commercial reactors to grasp the problem in its essence. These wastes will be deadly for thousands of years. The nature of the toxicity was understood from the beginning. Evolutionary prudence was ignored—and now the federal Department of Energy is planning on burying these wastes ("safely and permanently") for a period of time greater by many factors than the lifespan of all civilizations put together. The side effects, the "externalities," of technology have now entered the realm of geologic time. Yet all cries and pleas for eliminating radioactive waste at its source (weapons and nuclear power plants) are arrogantly ignored and rudely rejected as the irrational screams of misfits, primitives, and romantics. What we see here is "technique" at its most compulsive; nothing must stand in the way of progress and growth. Anything that does stand in the way is stupid at best and evil at worst. Backwardness is an intolerable drag on progress. Spiritual weakness, when hooked to technological arrogance, produces an increasingly blind and deadly monster that grimly refuses to repent of its follies and excesses precisely because its aggressiveness refuses to acknowledge its inner, ungrounded anxieties, its craving for ego immortality, and corresponding fear of death.

While Jacques Ellul provides us with an ironic analysis of technique, the independent radical Ivan Illich flatly states the cost of technocracy in terms of human freedom. "The more the citizen is trained in the consumption of packaged goods and services," says lllich in *Toward a History of Needs*, "the less effective he seems to become in shaping his environment." The more sophisticated and complex technology becomes, the more excluded ordinary people are from the control—or even the use—of such technology. Illich says that "a joyful life is one of constant meaningful intercourse with others in a meaningful environment."[7] This is of course true; but such meaningfulness is possible only where people are personally and cooperatively self-directing, and where sustained self-direction is in fact possible. We are back again to the conflict between personal fulfillment and cooperative culture on the one hand, and privatized advancement and rationalized civilization on the other. We are back to the tension between common culture and elite civilization and the extent to which the latter, industrialized, has totally overpowered the former.

IV

It may be possible to gain some additional clarity regarding cooperative culture and institutional civilization by attending to the words *friendship* and *peerage*. The two terms, and their respective implications, are so often intertwined that it may seem difficult to examine one without some partial view of the other. Peerage connotes layers of stratification, while friendship may blossom wherever there is courage and affection.

The first impulse to define peerage seems to manifest itself in terms of job or profession. Although one may have friends among one's peers, one's peers are not necessarily one's friends. Peers are those people whose position, rank, experience, or power roughly matches or rivals one's own. Familiarity is implied, but not, in the true sense, intimacy. Peers relate primarily to occupation or social position. They retain a formal or quasi-formal circumspection, even though interaction between or among them may be quite amiable. Peerage tends to relate to a formal, pre-existing structure; and it is appropriate to identify that formal structure as economic, political, or professional in nature.

At best, a professional peerage will maintain an explicit unity, a format of proper conduct that generates principles for its members and procedures for coming to the aid of a beleaguered member. (The traditional codes for university professors, lawyers, and doctors come immediately to mind, although these professional associations are now largely given over to creating protections and advantages for their own material advancement—or simply overwhelmed by university administration, giant law firms, or healthcare corporations. The "community of scholars" Paul Goodman used to write about can hardly be said to exist.)

At worst, a professional or quasi-professional peerage degenerates incrementally into *personnel*; this happens when the attendant organization becomes the primary and overriding reality, when professionals become only hirelings and slot-fillers in an organization managed by administrators. A serious question arises at this point: whether peerage in a managed corporate organization could be, by definition, professional. It would be difficult for anyone in such a context to have anything independently to *profess*. There may well be complex exercises to perform, including tasks that require great skill and extensive training; but there would be no room for genuine or independent originality.

There is no doubt an ambiguity here, for an association of true professionals—let us take a college faculty as an example—invariably is also an institution. The deciding factor then becomes a question of direction, policy contribution, and control. If a college *is* its faculty (and faculty-student relationship), then its profess-ors are not the organization's personnel. Personnel are hirelings and slot-fillers in an organization in which they have little or no policy voice; they probably don't know who or what controls the corporate body, and perhaps don't know or even care about the purpose or outcome of their activities. Soldiers are of course the prime model for the designation *personnel*: institutional killers and trained bombfodder who must be emotionally brutalized before they will "voluntarily" engage in rational slaughter. But increasingly all employees within large bureaucratic institutions have been infected by the spiritually toxic disease of *personnel*. Even the traditional professions have not been immune. As Ivan Illich has said, "Educational agencies that employ teachers to perform the corporate intent of their boards are instruments for the depersonalization of intimate relations."[8] The sociologist Lewis Yablonsky, in his book *Robopaths: People as Machines*, states the issue even more categorically: "People in a bureaucratic social machine do not give or take orders as people, but only as 'statuses' or as positions in an impersonal, ordered *social machine*."[9] Andre Gorz, in *Ecology as Politics*, says "The institutional function which has been passed on to the school is to perpetuate and confirm—not to counter or correct—the disintegrating, infantilizing, and deculturing action of society and the state":

> Schools do not teach us how to speak foreign languages (or even our own, for that matter), how to sing or use our hands and feet, how to eat properly, how to cope with the intricacies of bureaucratic institutions, how to look after children or take care of sick

people. If people do not sing any more but buy millions of records to have professionals sing for them, if they don't know how to nourish themselves but pay doctors and the pharmaceutical industry to treat the symptoms of an improper diet, if they don't know how to raise children but only how to hire the services of childcare specialists 'certified by the state,' if they don't know how to repair a radio or fix a leaky faucet or take care of a strained ankle or cure a cold without drugs or grow a vegetable garden, etc., it is because the unacknowledged mission of the school is to provide industry, commerce, the established professions, and the state with workers, consumers, patients, and clients willing to accept the roles assigned to them.[10]

So we can see, within the word *peerage*, a wide spectrum of possible meanings, ranging from mutual recognition and self-regulation to a kind of totalitarian uniformity—even if this uniformity is splintered into countless roles and specializations, and covered with a veneer of bureaucratic affability. Although the reality of friendship need not become manifest anywhere in this peerage spectrum, it seems reasonable to assume its most likely appearance would be on the self-regulated end of the scale. True friendship, unless it be the necessarily deep and trustful friendship of co-conspirators seeking to survive in a soulless corporate wasteland, would be practically impossible under the conditions in which personnel proliferate.

Writing before the First World War, the German sociologist Max Weber saw that bureaucracy sought to eliminate from the routines of its official business all emotional or irrational elements, all that was purely personal, all vestiges of passionate love or hate. Weber said that bureaucracy was a unique and peculiar form of organizational machine. Lewis Yablonsky endorses this perception: "Bureaucracy is an ultimate system wherein people's humanism is consciously reduced to a minimum level. Human emotions and expressions are antithetical to the smooth functioning of a bureaucratic social machine."[11]

When bureaucracy overtakes culture, people become personnel. If we take seriously E. F. Schumacher's assertion in his *Small is Beautiful* that the ideal of industry is the elimination of the living factor, and if we take note of the growing tendency to include under the term "industry" many traditional human activities formerly considered cultural, then the trend toward the industrialization of living culture becomes unmistakable. (I have in mind here such varied arts as education, healthcare, agriculture, and recreation.) It is in this context that the debate over technology takes on substantial meaning. As Andre Gorz puts it, "The struggle for different technologies is essential to the struggle for a different society . . . The total domination of nature inevitably entails a domination of people by the techniques of domination."[12] One might paraphrase Schumacher by saying that the ideal of industry is the production of personnel.

Recently the local school district here in northern Wisconsin hired a new superintendent who, in his remarks to a newspaper reporter, talked expansively of his "role" in the "expanding education industry." In light of the prevailing compulsion to industrialize culture, one waits patiently to hear or read about the industry of religion.

ENDNOTES

1. Mander, *In the Absence*, 178, 182, 186–87.
2. Mumford, *Myth*, 174.
3. Zuckerman, *Nuclear*, 105-6.
4. Ellul, *Technological*, 340.
5. Ellul, *Technological*, 134.
6. Ellul, *Technological*, 340.
7. Illich, *Toward*, 56, 91.
8. Illich, *Toward*, 77.
9. Yablonsky, *Robopaths*, 70.
10. Gorz, *Ecology*, 35.
11. Yablonsky, *Robopaths*, 69-70.
12. Gorz, *Ecology*, 19, 20.

9

The Industry of Religion

THE IDEA OF CALLING institutionalized religion an industry raises some interesting possibilities. In fundamentalist religion, there is a clear and unhesitating assertion that the entirety of life, the whole of nature, is the product of "God's Plan." One hears talk of "God's Blueprint." In early Deist thought, God was cast as Master Mechanic of the clockwork universe who, although retired from the present scene, was still Designer in Chief. In all these conceptions, the Creator is depicted as the Cosmic Developer with a surpassingly rational male mind. This is a God of awesome transcendent power, outside of and beyond nature.

Intellectually, it is a short step from a theological blueprint to a bureaucratic organizational chart. Jerry Mander paints a very clear picture:

> Judeo-Christian religions are a model of hierarchical structure: one God above all, certain humans above other humans, and humans over nature. Political and economic systems are similarly arranged: Organized along rigid hierarchical lines, all of nature's resources are regarded only in terms of how they serve the one god—the god of growth and expansion. In this way, all of these systems are *missionary*; they are into dominance. And through their mutual collusion, they form a seamless web around our lives. They are the creators and enforcers of our beliefs. We live inside these forms, are imbued with them, and they justify our behaviors. In turn, we believe in their viability and superiority largely because they prove effective: They gain power.[1]

Bureaucracy replicates what is allegedly God's primal pattern or impulse: a thorough ordering of society from the top down, with sovereign power in charge and in control. It seems credible to suggest, therefore, that mechanistic images of society have at least some of their roots (if it can be said that mechanistic images have roots) in religious conceptions. Certainly not all religious ideas or images are mechanistic; but let us consider what excessive rationalization has done to religious sensibility, and let us do so by attending to the thought of some additional weighty thinkers. First, from *Man and Technics*, Oswald Spengler:

> But it is of the tragedy of the time that this unfettered human thought can no longer grasp its own consequences. Technics has become as esoteric as the higher mathematics which it uses, while physical theory has refined its intellectual abstractions from phenomena to such a pitch that (without clearly perceiving the fact) it has reached the pure foundations of human knowing. *The mechanization of the world* has entered a phase of highly dangerous over-tension.[2]

In behalf of "physical theory" and "intellectual abstractions," it is necessary to point out that in the many decades since Spengler penned his words there has arisen a feisty band of impeccably trained scientists who hold that Earth is a living being and that life itself is sacred. These scientists are, of course, regarded as the hippies and beatniks, the bohemians and animists, of what is loosely called the scientific community; but the sheer fact that such dissenting scientists exist and that their conclusions were arrived at through thought processes whose insights and equations are part and parcel of the scientific baggage suggests that science is too real and too large a matter to be entirely contained within the boundaries of the technocratic mind. But that the "mechanization of the world" can be characterized by "over-tension" is radically more true now than it was either immediately before or just after the First World War: consider the present common use of such terms as "globalization," "extinctions," "global warming," "climate change," and "weapons of mass destruction."

Let us move on, once again, to Jacques Ellul:

> Nothing belongs any longer to the realm of the gods or the supernatural. The individual who lives in the technical milieu knows very well that there is nothing spiritual anywhere. But man cannot live without the sacred. He therefore transfers his sense of the sacred to the very thing which has destroyed its former object: to technique itself.[3]

This is a very stark analysis. Historically, the Christian religion has been at best ambivalent in regard to the spiritual quality of the created world. Those within the orthodox religious traditions will agree that Earth once was good—before the sin of Eve and Adam, the Original Sin, resulted in a fallen world and a fallen nature. In the sixteenth century, Martin Luther went so far as to say the entire world was under the dominion of the Devil: the wicked world was ruled by pure evil. This left only God and the hope of salvation after death to those whose life could only be a rite of suffering. Take away, then, the hegemony of otherworldly religion, as has happened in the modern period as scientific explanation has replaced religious myth, and one is left with a spiritless, desacralized world. I mean here that even those people—and they are numerous—who hold tightly to religious conviction also live, as Ellul says, in a technical milieu with nothing spiritual anywhere. Mircea Eliade, the historian of religion, saw all this many years ago:

> The cosmic liturgy, the mystery of nature's participation in the Christological drama, has become inaccessible to Christians living in a modern city. Their religious experience is no longer open to the cosmos. In the last analysis, it is a strictly private experience; salvation is a problem that concerns man and his God; at most, man recognizes that he is responsible not only to God but also to history. But in these man-God-history relationships there is no place for the cosmos. From this it would appear that, even for a genuine Christian, the world is no longer felt as the work of God.[4]

In this statement by Eliade, it is the cosmos and the world that have become inaccessible to modern Christians. In the matrix man-God-history, there is no place left for cosmic awe and childlike wonder; the world is no longer felt to be the work of a Creator. (There is an important distinction here, and to grasp it is crucial. There are a great many people who say they *believe* that "God created the world," but this is by no means the same quality of understanding of those who *feel* the world as the work of the Creator or as Living Creation. This distinction is compounded and made more agonizingly complex by the fact that many *believers* in God-as-Creator have an overbearingly instrumentalist view of nature, while many who simply *feel* the world as sacred are shunned as—probably—wicked pagans, people whose spirituality is not to be trusted precisely because of its inclusion of tender feeling towards nature or creation. But belief is only mind deep, while faith sinks into the bones.)

In another passage in *The Sacred and the Profane*, Eliade gives an example of what he is talking about. He says the "gradual desacralization of the human dwelling" is part of a process of "gigantic transformation made possible by the desacralization of the cosmos accomplished by scientific thought and above all by the sensational discoveries of physics and chemistry." He goes on to point out that the French architect Le Corbusier carried this trend to its logical conclusion by designing a house that is "a machine to live in."[5] It takes no great knowledge of history to see that this process of desacralization has occurred simultaneously with the enforced deterioration of folk culture and rural society, for folk society was civilization's cultural point of sustained contact with nature. Utopian civilization is in the process of crucifying both folk culture and global ecology. For civilization, with religion on its side, to destroy its cultural contact with nature results in the deification of technique, gross and growing insensitivity toward Earth, and an accelerating understanding of the spiritual that must be strictly (and even desperately) otherworldly.

What we in the industrial West call progress derives at least in part from the idea of heaven. As religion has lost its depth of power over society, and as society as a whole has become secularized, the idea of heaven has not simply disappeared. Rather, it has become transformed into the technological artifacts that science has invented and into the images that advertising uses to lure and tantalize us. Like the notion of heaven, both technology and advertising promise us a liberated tomorrow. But this "liberation" comes with a price. Getting to heaven requires an ongoing self-repression; getting the goodies that advertising dangles requires staying at the same old meaningless job. Technology, advertising, and heaven do not satisfy the ancient need to live in a world of felt cosmic awe. All of them agitate this need, and all frustrate it.

To feel the presence and power of the Creator behind or within creation is to feel the pull of immanent eternity. This tug of the eternal may be envisioned as occurring or developing in the future, but it is an *earthly* eternity. It is an eternity we live in and with. Rather than a world that comes to us from without (either by salvation or by purchase), eternity comes to us from within. The kingdom of God is *within* us, *among* us. It is, as Jesus said, "at hand." The inner and outer unfold in a living tension of harmonies. This is an eternity that is therefore here and now. (If eternity is all-inclusive time, how can it ever be other than always here and now?)

We might call heaven a deferred and sublimated form of eternity—not now, but later. This sublimation immediately reduces (it practically eliminates) the ethical tension that would otherwise compel us to live within the limits of nature or to feel nature as sacred and living. But if heaven is a sublimated form of eternity, then progress may be, in part, a desublimated form of heaven. That is, progress does have to do with real and tangible things; it has to do with a technological ideal of improving the quality or at least the tools and equipment of human life. But progress, as we are here talking about it, is tied into a world-view of scientific objectivity, a world of dead and fallen nature, a world of instrumentalist economics. This scientific objectivity is, historically, the product of highly civilized European minds whose thought was shaped by skeptical humanism. But even this humanism had Christianity in its intellectual ancestry. Humanism therefore sought to bring heaven down to Earth through civilized progress. But how does progress know when it "gets to heaven"? It cannot know because its core impulse is civilized aggression rather than spiritual humility, and so is always going and never there. But it also is always there, for eternity is here and now, and thus progress destroys Earth for the sake of an ideal image of technological perfection. Ellul has said that we cannot live "without the sacred." I would like to amend that by saying we cannot attain ecological sensitivity without the *feeling* of an awesome, living creation. Insofar as this earthly eternity is scorned or simply neglected by institutional religious practice, our need for eternity returns in the distorted garb of progress and transfers itself "to the very thing which has destroyed its former object: to technique itself." As Ronald Reagan used to tell us when he sold refrigerators for General Electric, "Progress is our most important product."

It is not for nothing that fundamentalist religion tends to applaud, support, and uphold the principles of progress, economic growth, and military prowess. Such a linkage derives from earthly sacredness denied. Since God is supposedly distant and withdrawn, man must impose civilized order on an unruly world. (It is a stunning paradox that explicit fundamentalism insists on a literal interpretation of the opening chapters of Genesis and yet is so coldly insensitive to the actual, physical creation.) Refusing to acknowledge ourselves as living within nature and the natural community, we armor ourselves against nature and fantasize an immortality of eternal escape—an otherworldly immortality gained through the rigorous denial of natural value and bodily desire, an imitation immortality purchased with the commodities promoted by advertising. Therefore, religion not only stands idly by while secular technology debases Earth, it becomes more and more manic in regard to End Times anticipations. Only a renewed sense of community in touch with the cosmic sacred can

burst through this immobilizing religious paralysis, this religious catatonia whose Armageddon fantasies reveal the explosive inner blockage
and turmoil. In the context of global technological civilization, only the
renewal of earthly sacredness can prevent the devastation of Earth.

II

Our estrangement from the cosmos, our transference of the sacred from
creation to technique, and our participation in the dangerous mechanical
overtension in the world will not be resolved by more rigorous church
attendance or by any form of revived religiosity in the conventional sense.
Going to church now means little more than transporting oneself or one's
family to the local organizational headquarters for professional worship.
In order to enter respectfully into God's inner sanctum, it is necessary to
put on a business suit and tie or a smart dress and new shoes—clothes
that are functionally useless and that, by being worn on this highest of
occasions, deny the value and integrity of one's hardy work clothes and
the physical life they represent. People who work for a living dress like
the business class when they go to church. These fancy duds are the finery
in which one is destined to be buried, perhaps in order to be in some
semblance of respectable civility when one arrives at the gates of heaven.
One hollers at the kids to hurry because it's already late; we jump in the
car and rush to that odd (and extraordinarily expensive) building that
otherwise has no purpose or function in our lives; we clamor up the steps
and sit breathlessly with countless others, mostly unknown, all passive,
who have similarly arrived, people with whom we have little or no sustained relationship. We are shouted at for being bad people, doomed to
hell unless we believe in God and capitalism, and we pay handsomely for
the performance. The preacher is sincerely acting out his role as the moral
barometer of civilized values. All this, in its institutionalized alienation,
is part of what *sustains* our estrangement from a more fulfilling personal
and communal life, from a more coherent culture, and even from a more
deeply felt cosmic sacredness.

The contemporary church is, largely, an organizational husk from
which the Creator Spirit has fled in boredom and in spiritual self-defense.
Many people, apparently, are becoming aware that something is seriously
wrong in the church. They improvise, modify, loosen up, or become more
grimly fervent. But they don't break loose; they stick with undiluted tran-

scendence. They try to make church "relevant." But because Earth is only a stepping stone on the way to heaven (as wheat is only a raw material in the splendor of white bread), "relevance" itself becomes another slippery abstraction. The core problem—ecclesiastical organization as structural impediment in the way of organic community—remains outside the realm of considered possibility. The world seems doomed, and one begins to yearn for the Final End simply to relieve anxiety.

Within the institutional mind-web, the buzzing of the constrained emotions becomes more frantic or resigned; desperation or inertia builds. Religion becomes even more abstract and otherworldly; daily life drifts further and further from the communal potentials of active faith; the credibility gap between religious assertion and daily life widens and expands. Yet the religious solution to this dilemma is to make religion even more private than it already is, to become exclusively salvation oriented, to avoid any hope or yearning—indeed, to reject and ridicule such yearning—for a fulfilling life in *this* sacred cosmic world. Such rejection and ridicule lead alarmingly to a religious-based desire to expedite global conflict and political disaster because such catastrophe may hasten End Times and usher in eternal relief.

Faith remains mere belief until the shackles of organization are broken and discarded. The "belief system" of a religious ideology is by no means identical with *faith*. Belief is a function of the rational mind, while faith involves the entire self, the whole body. An emphasis on belief leads to the pre-eminence of ideological "fundamentals." If one is saved by belief, then to have the *right* belief is a concern of the highest importance. Conversely, preoccupation with right belief places high value on the right attitudes of mind: the mind summons eternal vigilance in its babysitting of the inherently wayward body. Such wayward bodies are, in the collective, as Robert Rodale said of weeds, nature's unruly mob, entities to be uprooted in the cultivation of God's straight and narrow, or sprayed with divine chemicals that poison the mob but spare the elect. The mind and its right beliefs link tightly to the expectation of a heavenly future. The body is only a temporary encumbrance, and its mortal fate stands as a metaphor for the fate of Earth.

All this reflects belief, not faith. Faith is akin to what Buddhists call *satori*, a deep enlightenment that fills the body and alters consciousness. To those people for whom faith, not belief, is the bedrock of spirituality, eternity is not locked in the closet of the future. You needn't die first to get there. Eternity is the whole of time, the whole of space. Whatever else it may be, eternity is also here and now. This knowing raises the body and Earth to new levels of meaning and understanding. Only then does one begin to realize how belief, by asserting itself as the policeman of the body and prison guard of the backward, can with complete conviction and utter sincerity act on its own, without the presence of faith. Belief comes, thusly, to be the oppressor of faith, just as the church, as an institution, acts as the oppressor of faithful community.

People who should know better claim that "church is the only community we have," and they fear sustained isolation would result from the contraction of the church's organizational structure. These people fail to realize that the isolation created by belief is what the formal structure feeds on, and that community would be spontaneously generated (it is in the nature of human association) once enough people had overcome their structural addictions and anxious withdrawal symptoms. That we cannot maintain our belief-ridden egos in eternal perpetuity is something of an adolescent trauma every adult has to resolve in existential solitude and struggle. Lust for heaven is idolatry of the ego. It's all about protecting and perpetuating what's me and mine. It is anxiety encased in fear. Faith arrives as we let go of belief.

This line of thought may seem to be stretching social criticism to places where it does not belong or easily rest—where only fools rush in and angels fear to tread. After all, we are told, salvation is a matter between the individual and God; Earth, human community, and how one lives on Earth do not enter into the spiritual equation. Salvation has to do with a "free choice," and, like property, is strictly private. This world is going to pass away; it is only the Next World that should concern us.

It may be that if this religious viewpoint was held by people whose lives were ecologically prudent, we would think it outstanding. But such is not the case. Public concern over safe entry into the Next World only serves as a smokescreen for limitless technological meddling, gross industrial expansion, and military adventurism: a fallen world not only may be exploited at will, its atheistic troublemakers and pagan degenerates deserve to be pushed around a little. This is why the political Right links so easily with the religious Right: religion articulates a "dominion" justification while industry exploits the fallen Earth, and the military guards them both.

It is not those who hanker after community and an unspoiled Earth who are afraid of or opposed to the mysterious purposes of life. It is, rather, transcendent hubris that tramples the unspectacular immanent underfoot in its aggressive, frantic search for the immortal and the monumental. Technology seeks by industrial assault, under the disguise of progress, to attain a kind of abstract immortality, a technological heaven, a hopeless and wildly destructive illusion. This restless search, this powerful rootlessness, has now by means of accrued technology a destructive power that staggers the imagination and threatens to annihilate life on the planet. Cosmic awe, the small community, the unarmored self, subsistence livelihood, organic cultivation—all seem so trivial and helpless by comparison. Let those who seek this unpretentiousness be called, with sound spiritual justification, the meek. Let us have faith that Earth is still the promise given.

The organized church has become a husk. It has essentially fulfilled its historic mission: the factual spreading of the gospel. (Though here too it is necessary to stress that the gospel—if one means by that term the elemental teachings and parables of Jesus, along with his primary proclamation of the kingdom of God—has been only a small part, perhaps a tiny part, of what the church has taught. Go to almost any church service, pay close attention, and you will be astonished by the mere slivers of gospel and the huge slices of doctrine and creed recitation.) Only by the recovery of its "primitive" vision—the simple teachings of Jesus and the common, shared life of the early gatherings—could Christianity contribute substantially to the formulation and formation of a new social spirit. By freeing its energy from organization, the Christian church could help revitalize spiritual life through community. Organization and community are radically different spiritual realities. Organization devours community and excretes it as personnel. Excessive organization transforms what had been wholesome folk conditions into utopian civilizational structures, structures both ideal and empty.

True community is the inclusiveness of whole lives in an ongoing cultural continuity. The life of faith and community must be found elsewhere than in the dressed-up fragments of a weekend morning. Such life can still be rediscovered in the common work of a subsistence life close to Earth. Churches and church members are frozen in their inert patterns of expectation and denial. Church inertia is as great as the inertia of any other organizational system. The church is one of the interlocking institutions in the privatization and denaturalization of society. The church, we might even say, using its own apocalyptic terminology, is also the Beast. In fact, the church as presently construed is one of the cornerstones of abstract evil, for it claims to know Right from Wrong and the Way of Salvation—and it claims an *exclusive* knowledge. It claims Christlikeness. It claims to uphold, support, proclaim, and practice the earthy teachings of Jesus, sharing a life of suffering with the poor, outcast, and sick. But its organizational predilections, like its architecture, belie and contradict its pious assertions. By mouthing the text of faith in the service of belief, the church makes faith appear inherently smug, self-serving, hypocritical, foolish, false, and crass; it casts its abstractions against natural truth. It crushes the joyful subsistence of faith under the fretful commodities of belief.

The Creator Spirit moves without our permission or consent. The Spirit is not confined to a stone, glass, or wooden cube, frozen on a street corner, waiting for an infusion of warm bodies to thaw it out or a bolt from heaven to jolt it into alertness. The Spirit is a river that flows out of its banks. Organization may provide us with the delivery of technological progress, but only community can help us along toward spiritual fulfillment. And fulfillment, as the novelist Frank Waters has stated in his *People of the Valley*, is "individual evolution. It requires time and patience. Progress, in haste to move mass, admits neither . . . For faith is not a concept. It is not a form. It is a baptism in the one living mystery of ever-flowing life, and it must be renewed as life itself is ever renewed."[6]

It is a disturbing paradox that those churches that focus heavily on biblical apocalypse and who hold that, in the Last Days, the totalitarian organizational structure of the Antichrist will restrict all legal monetary and commodity transactions to those authorized personnel who have "the mark of the Beast"—those folks are generally in the front ranks of those who promote private property, capitalism, and massive national "defense." Ferociously anticommunal, these fundamentalists seem to be ignorant of or indifferent to the implications of their own apocalyptic message: for if, in the social uniformity of the computer age, only those who have "the mark of the Beast" (let's say an all-purpose plastic card or electronic implantation that serves as legal identification, credit transaction, library card, driver's license, and international passport)—if only those people who have this card or implantation can buy or sell or move about freely, then one of the means by which to identify the Beast or the Antichrist is authoritarian centralization, whether or not this centralized power has garbed itself in Christian pieties, but especially if it has. (How many years will it be before we begin to hear from the pulpits of fundamentalists that democracy is a failed experiment in "secular humanism" or that evil terrorism necessitates a "Christian" electronic implantation?)

In other words, to be outside this centralized commodity-saturated and heaven-ridden system requires precisely what the religionists scorn: earthy subsistence community. This startling contradiction raises the possibility that the organized Christian churches, standing simultaneously for otherworldly heaven and technological progress, are themselves "prophets" of the Antichrist: their apocalyptic expectations of global disaster are self-fulfilling prophesies, projections of their bleak and rigid inner emptiness, a consequence of their relentless "dominion." It further suggests that the husks of Christendom will seek in the end to suppress and exterminate the very Spirit they claim to be upholding and advancing. Should we really be so astonished that the organizationally "mature" Christian church, in which the ideology of belief has suppressed the cosmology of faith, would turn out to be the killer of the spirit of Jesus—the earthy holy man who founded no organization but whose life was taken by centralized organizations (both Jewish and Roman) threatened by his success in raising spiritual consciousness and transforming the body of human community?

In the Judeo-Christian creation myth, there were two trees in the Garden of Eden whose fruit Eve and Adam were not to eat. The first, of course, was the Tree of Knowledge of Good and Evil, and it was of this tree's fruit that they ate. But Adam and Eve were thrown out of Eden only incidentally for eating what they did: the larger problem was that they might, by lusting after immortality, eat of the Tree of Life—and live forever. It has taken a long time to realize that the roots of knowledge are deeply entangled with the roots of life; and, incrementally, scientific knowledge seeks to gain technological control over natural life and use it in privatized and even patented ways. This quest for scientific mastery is closely tied, in my opinion, to the same psychological dynamic that results in the exultation of belief. Conversely, the Tree of Life is rooted in the quiet, confident, sometimes agonized soil of faith. The fruit of the Tree of Knowledge is now being used to interrogate and intimidate the Tree of Life, and it is being done in the names of Science and Progress, the secular offspring of civilized, religious belief. We are once again at the gates of Eden—only this time with a search warrant and a van full of technical apparatus. This time the technicians of belief will not be inclined to stop until they have performed all possible experiments in genetic engineering on the Tree of Life.

If community "comes, like life, without machinery or artifice," as Baker Brownell has said, then it is our task in this exceedingly dangerous age to pick our way very carefully through the machinery of technological civilization and the slippery artifice of religious ideology. Remember, too, what Brownell added: "Life under wholesome conditions has a way of assembling itself in a coherent pattern. It has what may be called *organic intelligence*." (Emphasis added.) It is precisely the organic intelligence of faith, the real fruit of the Tree of Life, the seeds of which were cast in the parables of Jesus, it is this organic faith that belief will neither understand nor tolerate. In the name of knowledge, life is abused. In the name of progress, subsistence is derided. In the name of belief, faith is scorned. In the name of Christ, the followers of Jesus are called pagan. In the name of heaven, Earth is treated as a cesspool.

ENDNOTES

1. Mander, *In the Absence*, 209–10.
2. Spengler, *Man*, 93.
3. Ellul, *Technological*, 143.
4. Eliade, *Sacred*, 179.
5. Eliade, *Sacred*, 50, 51.
6. Waters, *People*, 134, 177.

10

Redemption of the Past

DESPITE THE STATEMENTS I have made in regard to organized religion, it is neither my intent nor my desire to attack spiritual truth or impugn the impulse of religious faith. I am not a "materialist." My anger toward conventional religion emerges from a conviction that adherence to creedal belief obstructs, substitutes for, and smothers ethical faith. Asserting "Christian nation" status for the United States, the Christian Right votes overwhelmingly for state violence and corporate greed. In terms of gospel truth and spiritual discernment, something is radically wrong with the Christian Right. Mythological belief has driven ethical faith from the church.

I do not, however, reject out of hand the possibility of some life force within us surviving the grave. Reality is invariably more subtle and complex than we think. The sheer complexity of human emotionality has often led me to wonder how any of us could have become who we are, or our relationships so wonderfully entangled, in a single lifetime. Or, conversely, if all this complexity is the consequence of single lifetimes, then the sheer creative power of nature (or what we so blithely and casually call "God") is so awesome and magnificent that our personal quailings are only a distracting form of selfish faithlessness. But I also believe we have an inherent religious capacity to feel the sacredness of the cosmos and to express our wonder and awe with tender and insightful artfulness. Getting past our belief-ridden blockage to that faithful capacity is our essential task.

It is also my conviction that spiritual and material realities are paradoxically interwoven. (If, in the Christian religion, the Creator entered physically into the natural world in the person of Jesus in order to live, suffer, die, and be resurrected, then the metaphorical and metamorphical implication of Jesus' transformation is the painful crucifixion and even-tual resurrection of nature, organic intelligence, and ecological community. And if all this is true, doesn't it follow that the current "globalization" process, with its elevation of the Seven Deadly Sins—pride, covetousness, lust, anger, gluttony, envy, and sloth—into commercial and military virtues, is facilitating that crucifixion? Resurrection, in turn, implies a coming back to life, life in a far deeper ethical capacity and spiritual realization, a life of humility, depth awareness, and earthly contentment.) When all is said and done, we hardly dare assert—short of poetry, music, dance, or mystical insight—who we are or why we are. Ashes to ashes, dust to dust. Awe, therefore, and unaffected humility are proper manifestations of our wonder as we meditate on our essential mysteriousness in the vast cosmic panorama, as each of us sees, hears, tastes, feels, and smells our daily way through this totally inexplicable conscious voyage in our fleshy crucible called human life.

We are approaching the end of a peculiar time in history when neither the vast cosmos nor our small Earth has been held as sacred—except, perhaps, by children, animists, and "primitives." There is no need to dwell on the material benefits generated by the programs of rational skepticism: the media that bring us its blandishments are bought and paid for by the advertising revenues of its commodities. Reform by advance has crushed traditional retrogression; but a new, more spiritually robust and enlightened retrogression is struggling to come alive and find its footing in the renewal of organic community, a fuller kind of culture at once more earthy and more ethical.

It would, however, be foolish to assert that skepticism has brought us nothing of value. It has brought us much; it has vastly broadened our world and our view of the world. But because civilization's underlying motives are so skewed and privatized, so power-driven and arrogant, the technology and economic structures of rationalism are correspondingly dangerous and destructive. They are instruments of profit-maximizing predation and power-drunk hubris. They threaten to destroy, in fact, the very basis of our earthly existence. It is not mere rhetoric to insist that unless we rediscover the sacrality of Earth and cosmos, and unless we do so with a rapidity and fullness that defies historical comparison, we might well perish with the poisoning of Earth.

Our connectedness to the natural world and to stable human communities can be reestablished, although only in a new cultural configuration, by fusing our spiritual values and sense of cosmic wonder with a life at once less dependent on institutional commodities (including prepack-aged religious commodities) and more vibrantly earthy. Such transfor-mation requires that we unhook from our overenergized comforts and securities—social addictions we have accumulated largely by allowing ourselves to become the institutionalized personnel of technological civi-lization—and begin to focus on the necessary steps by which we might renew durable bonds to the cosmos (fields and flowers, rivers and oceans, stars and animals) and committed relationships to specific people in our daily lives. The more saturated with artificial and abstract dependencies we allow ourselves to become, the more likely we are to lose touch with the age-old culture of communal interdependence and mutual aid. As Ivan Illich has put it, the "high quanta of energy degrade social relations just as inevitably as they destroy the physical milieu."[1] Ezra Mishan has stated the issue even more bluntly:

> [B]eyond a point in technological progress, and we are already be-yond it, the innate capacity of ordinary people for open and warm-hearted enjoyment of each other begins to shrink. Sublimation translates into alienation, specialization into disintegration.[2]

At the same time, it must be said that the development toward a freer and fuller individuality has not been without its real blessings; its most important fruit could well be a fresh and more equitable relationship between women and men—for traditional folk culture was grounded in gender-specific patterns of work and social conduct, and the breakdown of traditional folk culture holds, therefore, the promise of ungendered work in the creation of a new culture.[I]

Yet we must not permit ourselves to become trapped into either blind approval of or bland indifference toward those machinations and catas-trophes that have accompanied this very important but fragile growth in freedom. Expansionist, technological civilization has broadened the world by overpowering gender-bound, traditional folk culture; but civili-zation has its own sexist arrogance (the military as perpetual male fist, the civilized male mind as template of God's will) that, once supreme, makes equality a hollow and rotten fruit. We must hold to the visions of freedom and fulfillment; but we must realize that both freedom and fulfillment will

---

I. For a magnificently well-told story of breakdown in an African context, see Chinua Achebe's novel *Things Fall Apart*.

be crushed unless we learn—and quickly—to root them with reverence in ecological and communal coherence.

Civilization without the cultural coherence of noncivilized folk culture is destined to collapse of its own organizational weight and ecological stupidity. Freedom and fulfillment are not free rides on the civilized perpetual commodity machine. Unless we ground both freedom and fulfillment in common culture, the rapid deterioration of civilized infrastructure will make a mockery of our finest dreams and visions. We must take responsibility for our lives and culture by renewing our commitment to Earth and to community; we must break our dependency as passive consumers of the system's rapacious largesse.

Holding both to liberty and to equality, we can select appropriate reforms by advance and integrate them into our lives with adequate reforms by retrogression. By choosing the right paths, we can have material comfort *and* ecological integrity, personal fulfillment *and* communal cohesion. These are the paths of humility and reconciliation.

II

The facile technocratic optimism that suggests we will be able to proceed blithely down the broad avenue of industrial growth while living sleek and capricious space-age lives, comfortably removed from the dirt, toil, pain, and trouble of our everyday world, is a dangerous illusion. We are, as Carl Jung said, "very far from having finished completely with the Middle Ages, classical antiquity, and primitivity, as our modern psyches pretend." Only sustained ethical dedication to reform by retrogression will enable us to avoid another murderous outburst of uninhibited irrationality, an outburst that, given the available weaponry, could well be the last.

Mircea Eliade has shown that in the archaic past there never was the idea or the fear of time's *end* in the modern linear sense. History was not a linear progression. History was not a linear but a cyclical myth. That is, history was a never-ending wheel of life and death that turned one steadily around the "objective" cosmos of one's "subjective" awareness. This cosmos had (and still has) the rhythms of the seasons, the rising and setting arc of sun and moon, the seasonal wanderings and migrations of insects, birds, reptiles, and mammals, the cyclical coloration of Earth, the seasonal shifts in weather and storm, the movements of stars and planets in the heavens—all this and more for its cosmic calendar.

Only within the past few decades has it been demonstrated (by Gerald Hawkins, an astronomer at the Smithsonian Astrophysical Observatory, in a book entitled *Stonehenge Decoded*) that the old and mysterious Stonehenge in England is nothing less than an astronomical observatory whose very structure constitutes the embodiment of a whole matrix of mathematical equations: this huge stone circle, some of whose individual members are estimated to weigh thirty-five tons, was constructed by people believed to have lived nearly four thousand years ago. Stonehenge itself is so complicated a construction that it required advanced technology to unravel its purpose. So impressed is Hawkins with the astronomical intricacy of Stonehenge, he calls it a "Neolithic computer."[3]

It may be argued that Stonehenge is an exceptional oddity. Perhaps, and unredundantly. But it is more true that some assertions regarding the past—that we have nothing to learn from traditional patterns of life; that the human experience, to use Thomas Hobbes' classic deprecation, was invariably poor, nasty, brutish, and short—are merely indicative of a civilized arrogance that has the gracelessness to scorn the folk heritage as culturally contemptible. Perhaps we only reveal our civilized insensitivity when we allow our imperious affluent satiety to make crude and contemptuous allusion to the coherence and even splendor of the past. No less a radical than Herbert Marcuse has stated (in his *Eros and Civilization*) that "The past remains present; it is the very life of the spirit; what has been decides on what is. Freedom implies reconciliation—redemption of the past. If the past is just left behind and forgotten, there will be no end to destructive transgression."[4]

Crude or splendorous, the past is our own past; "primitives" are not some wholly unique race with whom our modern civilized lives have no relation. In the context of a human history that boasts hundreds of thousands if not millions of years of social and cultural evolution, what happened five or ten or fifteen thousand years ago is only a kind of yesterday. To think we can "escape" the past is absurd; we can no more escape the past than we can escape the needs and impulses of our inner being. And, indeed, there is a special bonding between the past and our inner needs, for our most basic needs were nurtured in the womb of the past and therefore find little satisfaction in what is new. In my opinion, those who are most fascinated with the latest novelty are generally those who are most out of touch with their inner needs, whose sense of the "present" is alarmingly trivial and shallow. Conversely, those who are most contemptuous of the past are often those who fight their inner needs as if those needs were evil demons. There is no *escape* from the past unless we literally blow ourselves up. But we can, as Marcuse insists, *reconcile* with the past and, by so doing, discover an unanticipated inner liberation. Those who would block the past are most vulnerable to repeat the follies and disasters of the past. Avoidance generates unconscious repetition. Armoring against the past only means, finally, that we will turn our weapons on ourselves.

Respect for the past is the key to how we approach reform by advance—whether specific projects are socially necessary, ecologically sound, and "convivial." (Ivan Illich has used the term to indicate, for instance, tools or modes of production that enhance rather than eviscerate the well-being of small communities. As Illich says, a growing "dependence on mass-produced goods and services gradually erodes the conditions for a convivial life."[5]) Reform by retrogression would have to accommodate the same criteria as reform by advance. Yet the sifting and winnowing that has occurred naturally in history makes choosing the proper retrogressive crafts considerably less difficult. We know, for instance, what makes a tool feel good in the hands or that vegetables harvested fresh from the garden taste better. The real problem here is twofold. First, it's necessary to overcome our commodity bloatedness as industrial consumers and find the necessary will for the recovery of vital retrogressions; second, we must reconstruct such retrogressive recoveries in a way that overcomes and transforms the stratification along sexual, racial, and class lines with which subsistence culture was often embedded. I truly believe this liberation is now possible; it is the felt eutopian promise within the deeper reaches of democratic self-governance, which, in turn, in Christian terms, is surely linked to the egalitarian ethics of simplicity and sharing articulated so concisely in the stories and parables of Jesus. The past may be heavily charged with rage, pain, and all manner of emotional knots, but it never was universally poisonous, as our contemporary technocratic civility is. Only by dealing honorably and bravely with the past will we be able to avoid an utterly venomous future.

Shall we go through the vast array of modern inventions and technical achievements and begin the difficult task of sorting the wheat from the chaff? What will we do with nuclear wastes? Is it prudent to continue to produce the variety and quantity of chemicals now in use, spread on the land, flushed down the toilet, or deposited in toxic waste dumps? What will we decide about manufactured (and patented!) life forms that have come out of genetic engineering? What are the "lifestyle" implications for our culture when acid rain and fossil fuel emissions are radically reduced? How can we simultaneously save Earth from our economic abuse and ennoble our culture? These are questions that are, at present, either avoided or answered with cavalier arrogance by the technocratic establishment. We don't take them more seriously because our level of comfort supplies too much false security, even if such security is a temporary luxury. We have come to believe our civilized institutions are invulnerable, in control, and able to manage any crisis. We believe our capacity for management is within the certitudes of permanence. We believe God has mandated our technological dominion.

But it is necessary to carefully examine our past, and to recover, where possible, viable methods of living and working. It may well be that by adapting and improving on the methods and tools of the past, we shall be able to have an adequately modern cake and safely eat it, too. In considering the rural culture implications of reform by retrogression, we might, for instance, imagine country neighborhoods that blended small-scale private ownership with cooperative land trusts; farms with modest numbers of cows, goats, sheep, and pigs; chickens, ducks, turkeys, and other poultry; large gardens, ponds, orchards, and vineyards; fields well cared for and unpoisoned, their productive capacities neither artificially stimulated nor exceeded (some blend of traditional practices integrated with the work of Wes and Dana Jackson and mulched with *The One-Straw Revolution* of Masanobu Fukuoka); horses to pull wagons and buggies, to ride, to skid logs and firewood; small tractors for plowing (unless regenerative agriculture fully eliminates the need for plowing) and heavier field work; workshops with extensive hand tools as well as electric drills, saws, planers, lathes, and so on; forges, anvils, and welders; barns well-designed and small enough so that animals can be individually attended to and kept in a proper state of cleanliness and health; well-insulated houses, cabins, sheds, and shacks built with native materials, heated by sun and fire, electrified by wind and photovoltaics; greenhouses attached to virtually every house and barn; comely community buildings for meetings, dances, musical events, and common meals. We can broaden the vision to include locally established schools, markets, restaurants, inns, hostels, food cooperatives (both consumer and producer), wildlife preserves and parks; cooperative use and ownership of big equipment; a neighborhood clinic for the seriously ill who could not, for whatever reason, remain at home; a well-designed neighborhood home for those people, elderly and others, who wished to reside together; a quiet cemetery.

The vision is real and wholesome. The needs it speaks to are simultaneously ancient and alive. The skills, tools, and techniques are all available. There is sufficient land upon which such communities can flourish. "To cooperate, in the highest as well as the lowest sense," wrote Henry Thoreau in *Walden*, "means to get our living together."[6] Indeed. Everything is at hand except the most vital ingredient: the clear, spiritual will of ordinary people. What's missing is the realization that the shimmering future with which industrial progress titillates our consumerist belief can never be lived at the level of real community and earthy faith. What *can* be lived in community and faith is infinitely more real than the ersatz "culture" of civilized consumerism. We are mesmerized by illusions, and these illusions are held in place by fear of the past, fear of the future, and the thwarted energy of our inner, unmet needs. To have a living culture, we must be real and living people.

ENDNOTES

1. Illich, *Toward*, 111.
2. Mishan, "Wages," 85.
3. Hawkins, *Stonehenge*, 174.
4. Marcuse, *Eros*, 117.
5. Illich, *Toward,* vii.
6. Thoreau, *Walden*, 58–59.

11

Urbanitis

W<small>E HAVE JUST BEGUN</small> to reach that point in the transformation of consciousness where growing numbers of people are realizing the pressing need for clear and decisive change. Yet there is a great deal of confusion about what the alternatives really are. We piddle with things like recycling (though recycling and the mindfulness it takes to do it are fine), not realizing how removed we really are from true ecological living. Industrialism has assumed such massive proportions in our lives—from our everyday consumption of its commodities to our saturation with its institutional and electronic "culture"—that to look critically at the industrial agenda is to risk radical departure from normality. Industry has become civilization; and, semantically, civilization claims sanctity similar to patriotism and to God. By civilization we mean not only the great urbane monuments of the past but, more importantly, the force of progressive linear development toward a rational and ultimate end. Secular ideology, having discarded heaven as a probable destination, and being unwilling to embrace the subsistence mortality of cooperative earthiness, is therefore a dynamic, restless engine without a depot. It can only proclaim, in its shrill demented rationality, that progress is our most important product—even if this "progress" is now hardly more than an ideological or even mythological abstraction manifest in largely trivial consumer innovation.

While progress, rightly understood, neither can nor should be abandoned as a worthy principle, civilization as historical progression needs to share its exalted position with the archaic principle of cyclic renewal, with the stable wheel of life. We need a new synthesis of awareness whereby energy-intensive advances are not allowed to shoulder out nonaggressive traditions simply by force or might. Might not only does not make right; it does not necessarily make better. In short, we need to work deliberately and

106

consciously toward the recovery of specific modes of cultural life that have been disregarded and discarded in the acceleration of civilized progress.

Does this need for appropriate retrogression imply a "return" to primitivity as historically understood? The abandonment, for instance, of advanced procedures for medical treatment and a regression to chants and herbs? The question is specious. It is not necessary to demean the past in order to assert the efficacy of the present; in fact, we may not be able to sort through medical advances without the spiritual intelligence that comes with the deliberate recovery of chants and herbs. There clearly have been major medical advances—in immunology, surgery, dentistry, psychotherapy, and so on. Yet such advances are only a part of the picture. We have, on the one hand, the recent rediscovery of such practices as acupuncture and natural childbirth, while, on the other, we are faced with the astonishing paradox that our modern way of life *creates* new disorders. The growth of cancer rates is only one example.[1] According to Jacques Ellul, "The November 1960 issue of *Semaines Medicales de Paris*, on the basis of information contributed by 4,000 physicians from all over the world, offers a study of a new disease of great complexity which is brought on by modern city life and which might be called *urbanitis*."[1] In his book *Our Synthetic Environment,* Lewis Herber has said:

> The metropolis establishes the social standards of the entire country. Owing to its commanding economic and cultural position, it sets the pace of national life and establishes nearly all the criteria of national taste. Many distinctly urban forms of work and play have invaded the most remote rural areas of the United States, where they generate the same stresses in the villager and farmer as they do in the city dweller. The nature of agricultural work, moreover, is changing. As farming becomes increasingly industrialized, diversified physical work is reduced to a minimum by machines and one-crop agriculture. Although the farmer still pursues a less hurried way of life than his urban cousin, he is often beleaguered

I. Michael Pollan in his new book *In Defense of Food: An Eater's Manifesto*, on pages 85 through 89, relates an experiment conducted in the summer of 1982, in which "a group of ten middle-aged, overweight, and diabetic Aborigines" in Western Australia voluntarily returned to the bush for seven weeks, relying exclusively on foods they hunted or gathered. When the seven weeks were over, nutrition researcher Kerin O'Dea drew blood samples from the volunteers and "found striking improvements in virtually every measure of their health." Triglyceride levels had returned to normal. Omega-3 fatty acids had increased. Weight loss averaged just under eighteen pounds per person. The Aborigines' prior diet of civilized food was obviously deleterious.

by even greater economic problems. Both in the city, and on the land, a new type of man seems to be emerging. He is a nervous, excitable, and highly strained individual who is burdened by continual personal anxieties and mounting social insecurity.[2]

In a historic context in which earthy folk culture has been crushed by standardized industrial civilization, and where organic community has been replaced by organizational personnel, it is reasonable that Herber was able to identify the trend toward nervousness, excitability, and emotional strain with the "criteria of national taste" generated by the metropolis.

Although I do not wish to deal at length with the subject of cities, it is necessary to make a clarification in regard to Ellul's use of the unusual term "urbanitis." And there is probably no contemporary thinker more capable of making that clarification than that distinguished lover of cities, Lewis Mumford. In his book *The Urban Prospect*, Mumford says:

> When one translates into concrete terms the current talk about the increasing urbanization of the United States today, one must understand that sociologists are speaking loosely of people who are, in fact, disurbanized, who no longer live in cities, or enjoy, except as visitors or part-time occupants, the concentrated social advantages of the city: the face-to-face meetings, the cultural mixtures, the human challenges.[3]

It is important to realize that most of the traditional advantages of city life have become as diluted by commercialism as has the cultural life of the countryside. As anyone who has had personal exposure to both cities and the countryside can attest, the deterioration of cities is as pronounced as the rural barrenness. These developments are, in fact, products of a single cause—unchecked industrial growth and heedless technological progress. Ringing the dreary, dirty, dangerous, and decayed cities are interstate highway loops with their brazen imagescape of billboards, shopping malls accessible only by automobile, and thickly packed unimaginative subdivisions—all of which strangle the cultural vitality of the city and make the countryside inaccessible to nonmotorized city residents. It is less accurate to talk about urbanization than *sub*urbanization.

Suburbanization brings with it the shopping centers, grocery chains, quick food eateries, self-serve gas stations, and all the rest of the artifacts of the instant society. The city itself is reduced to fewer components: financial and transportation centers, high culture institutions like art museums and symphony halls, the prowling police, and the human abyss of the welfare tenement. The inhabitants of the city find themselves blocked off from the countryside, especially if they are poor, black, Asian, or Hispanic; and many of those people left on farms think of the city as the very pits of hell. (There is a strong tradition of this anticity perspective in rural populism, some of it justified and some of it simply racist.) Meanwhile, speculators and developers gobble up productive agricultural land and convert it either into suburban sterility or industrial "parks," bereft of both the human intensity of the city and the natural openness of the country-side. The loss of sense of place brought about by this enforced alienation begins to set the cultural tone for the entire country. Our lives have less and less to do with nature and the particulars of our immediate habitat but more and more to do with organizational systems and electronic amusements.

Narrowly, whatever people do is culture. It does not follow, however, that there is no distinction to be made between kinds and qualities of culture. The culture that a people indigenously create and embellish in the context of earthy subsistence must of necessity be complex, tough, and adaptable. It must fit their basic needs. Such people know their environment and are in constant contact with it. But once disrupted by outside technology, especially highly sophisticated industrial technology, a native culture is doomed to disintegrate. People lose their capacity to sustain their cultural activities. They lose their intimacy with Earth. There are countless examples in American history alone to prove the point.

Similarly, there is within Western civilization a long and honorable tradition of rural folk culture, a culture that had its own complex relationship to Earth. I speak here of the European peasantry and the peasant commons. This culture, too, although it was indigenous to the larger society that eventually produced industrialism, was destined to be destroyed. That is, the peasant cultures of Europe were part of an organic cultural evolution that first appears in the Neolithic. Those European folk cultures were part of an agricultural tradition in an enforced subsistence mode, due to the sustained extractions of an aristocratic landlord class, and they were prevented from evolving into a richer, more complex culture precisely because aristocratic extractions kept them impoverished and overawed. Civilization, including the aristocratic civilization of feudalism, was a hybrid predatory culture that existed through enforced exploitation and systematic expropriation. For thousands of years, civilization was a self-conscious, righteous parasite on peasant culture. As Will Durant says in *The Reformation*, "Civilization is a parasite on the man with the hoe."[4] The simultaneous emergence of democracy and industrial capitalism threw this clear peasant perception into ideological confusion: to "democratize" civilization through industrialization it has been necessary to cover over the thoroughly *un*democratic dismemberment of agrarian culture. Dazzled and numbed by the commodities of affluence, trinkets in exchange for life, this has become a democracy of things for a constituency of personnel.[II]

To break up these complex rural cultures may be described by technological apologists as merely the enforced substitution of an outmoded, primitive culture by an advanced and improved industrial culture. But is this necessarily so? We have already seen to what extent the "average" American family watches television—one significant example of cultural catatonia. When we compare this with records of the past—say Eric Sloane's *The Diary of an Early American Boy*—there is simply no comparison in cultural depth. It is amazing what children two centuries ago knew how to do. And if it can be said that there was a certain level of coercion in earlier child competence, we must also ask to what extent such coercion was the product of endemic *class* coercion within civilized economics—a coercive trickle-down effect—and also whether, in a truly ethical and ecological economy, child competence might be restored with new and richer freedoms.

One of the most important features in the destruction of indigenous folk culture is the loss of common craft. When people can no longer make do for themselves, when virtually everything they need or desire must be a purchased commodity, they become dependent on institutions outside their control to a degree that is culturally atrophying and politically dangerous. They become consumers, pure and simple—consumers not only of things but of ideas and ideologies. In losing craft, we lose culture. We

---

II. When the Soviet Union collapsed, there was much talk in the West of a "peace dividend," a turning from military to domestic spending, a "dividend" that never materialized. Similarly, looking back, one would have thought that "democratic revolution" would have enabled the peasantry to regain its long-thwarted cultural evolution. (This may have been the buried intent of Jefferson's "agrarian vision.") Instead, "democratic revolution" resulted in the virtual extermination of rural culture. "Democratic" *civilization* has not restored the cultural integrity of the agrarian village. Rather, it has indoctrinated industrial consumers with images and prospects of aristocratic opulence while reducing agriculture to chemically drugged industrial agribusiness.

lose self-directedness. To say that industrial commodity consumption and mass media entertainment are adequate substitutes for the craft and community of preindustrial folk culture is an absurd proposition. The imposition of commodity technology removes functional craft from the hands of ordinary people; it creates the social conditions in which culture itself becomes a commodity to be mass-produced and mass-consumed, subject to sudden engineering alterations in the service of engaging novelty and faddish obsolescence. In removing our ability to subsist, even marginally, outside the technocratic system, industrialism renders us easy prey to political manipulation, imperialist jingoism, and even a new and deadly form of fascism, bland and cheerfully innocent in its indifferent cruelty.

We need to study the extent to which technology has become not only the tool of demented ideology but its foremost expression. Consider these remarks from the sociologist Philip Slater, as found in *The Pursuit of Loneliness*: "We turn continually to technology to save us from having to cooperate with each other. Technology … serves to preserve and maintain the competitive patterns and render it ever more frantic, thus making cooperation at once more urgent and more difficult."[5] To learn to cooperate at this stage of our civilized alienation means to rediscover our common culture and, by building it anew, reclaim it with our lives. This is at least one possible meaning of the word "resurrection."

II

In writing about the prospects for cultural renewal, one feels constantly up against the power of the industrial consensus, almost as if one spoke a language few understood and that was, for many people, a diffuse source of sardonic humor or simple embarrassment. One thinks, for example, of those people who dare to call themselves, in this day and age, Jeffersonians. True, the term has in time overgrown the complex ideology of Jefferson himself; but to be a Jeffersonian is to have grave misgivings in regard to the urban-industrial ascendancy. To be a Jeffersonian is to hope, however inchoately, for the reconstruction of a decentralized, rural stability. Because such views are for the most part held by people who feel unable to defend them adequately against sophisticated political opponents, it is refreshing to find an exceptionally well-regarded historian who also holds such "backward" views. We have such a person in the sagely Lewis Mumford. In *The Urban Prospect* Mumford says:

Though Thomas Jefferson's fears for the physical and moral health of his country if its predominantly rural culture became urbanized and industrialized have long been justified by irrefutable statistical evidence, they are still too often treated by historians as a pathetic bucolic prejudice ... So, far from looking to a scientifically oriented technology to solve our problems, we must realize that this highly sophisticated dehumanized technology itself now produces some of our most vexatious problems, including the unemployment of the unskilled.[6]

Certainly one of the key semantic distortions that has permitted technological apologists to get away with their mocking condescension toward rural culture is the term "industry." Once this word has been successfully applied to agriculture, the essential distinction between culture and technology becomes clouded and confused. In the Introduction to Lewis Herber's *Our Synthetic Environment*, William A. Albrecht, Professor Emeritus of Soils at the University of Missouri, addresses this issue:

Actually, agriculture is quite the opposite of industry. Good agriculture involves a cooperative enterprise with nature, indeed a series of creative operations that center on the living fertility and organic matter of the soil. By exploiting the soil exclusively for economic gain, we are pushing all the life forms of the earth closer to their ecological fringes of survival. So far as man is concerned, the consequences of this tendency take the form of a decline in the nutritional quality of food staples and forage crops and an increase in diseases and pest infestations. Instead of trying to remedy the situation by restoring the balanced relationships that exist in the natural world, we normally resort to methods and devices that have been developed for increasing sales profits.[7]

Urban-centered industry, in its pursuit of profit, power, and progress, has crushed preindustrial culture and substituted a thin veneer of sophisticated distraction it dares to call "high," "enlightened," and "popular" culture. If everything people do is culture, then this too is culture. But industrial culture, as the froth of the civilized system, has no organic depth; civilization has always existed by compulsory exploitation; and industrial civilization, whether or not we call it democratic, merely compounds the depth and degree of alienation. Since industrial culture has no depth, it must be made continually novel and sensational; it must activate our depth needs and impulses without satisfying them; it must keep us craving those things that do not fulfill the need that its advertising images activate. All this is kept as close as possible to patriotism and God; to be opposed to progress is to be a pagan, a heathen, or an uncivilized person. Religion is called upon, with its language of "dominion," to defend "man's mastery over nature." But all this can continue only on the basis of severe environmental degradation and intense energy consumption. The distractions that are called "culture" are not allowed to lessen for fear of civil disorder. Caught in this system without clear understanding, people experience chronic nervousness, excitability, and emotional strain. This is the price of buying into the "criteria of national taste." The dis-ease is urbanitis.

ENDNOTES

1. Ellul, *Technological*, 331–32.
2. Herber, *Our*, 68–69.
3. Mumford, *Urban*, 229.
4. Durant, *Reformation*, 752.
5. Slater, *Pursuit*, 133.
6. Mumford, *Urban*, 232, 246.
7. Albrecht, "Introduction," xi–xii.

12

Industrial Agriculture

To SOME READERS IT may seem that, like a modern primitive who has just reinvented the wheel (or wheelbarrow), my preoccupations with small-scale, communal, and organic agriculture are of a similar retrograde nature—something impulsively backward. But perhaps there is treasure in the barnyard mire; for it may be, as sparrows know of horses, there are whole grain truths that sometimes pass through the larger system undigested. It therefore seems a promising task to dig in the compost pile of history to see if we can find that which, though excreted and discarded by the body politic, might yet provide the humus for a new and richer field of culture.

To minds in the sterile mainstream, this is surely crackpot thinking. But in his Introduction to Helen and Scott Nearing's *Living the Good Life*, Paul Goodman explores the potential of just such "crackpot" ideas:

> [I]t is clear that we have to take seriously the Thirties' ideas of the Nearings, Borsodi, Frank Lloyd Wright, and the Southern Regionalists—and the economic ideas of Gandhi before them and, of course, the kibbutzim. Their experiments and analyses used to seem cranky, if not crackpot, and they were certainly not in the mainstream of the technical and political issues that were discussed. But suddenly we have reached a tipping point. Ecologically, we are facing disaster, both environmentally because of pollution and physiologically because of poisoning. Abuses of technology have gone so far so fast, that the chief present purpose of technology must be to try to remedy the effects of past technology. Everywhere in the world the galloping urbanization is proving to be ecologically and fiscally unviable; worse, it is impossible to bring up citizens in urban and suburban areas that are no longer cities. The processing and social engineering that go with these conditions have called forth waves of populist protest, articulate

and inarticulate, by those who are pushed around and find themselves without power. And, finally, the expanding Gross National
Product, the ever higher Standard of Living, which was the justification for all this, has begun to yield sharply diminishing returns,
trivial goods, incompetent services, base culture, and spiraling
inflation.

Thus, the eccentric ideas of the Nearings and the others are no
longer out in left field. History, alas, has caught up with them. With
a few more years of power failures, transit strikes, epidemics of
heroin overdose, water shortage, unacceptable levels of air pollution, crashing aeroplanes, hundreds of thousands of New Yorkers
will regard Scott and Helen as uncanny prophets. My own opinion
is that American society would be far more viable if we could push
the present five percent rural ratio back to something like twenty
percent, as an option and a standard of people who respect the
environment and who, as Jefferson pointed out, cannot be pushed
around because they can feed their faces.[1]

Goodman wrote this Introduction around 1970, when farm population was still close to five percent. To his proposal for a twenty percent
"rural ratio," I would only suggest that twenty percent may be too modest,
although it would be a serious and significant improvement over the present condition.

A more densely populated countryside could generate a good deal
more cooperative self-provisioning and a great deal less reliance on
chemicals as poisons and fertilizers. This is partly a function of small-scale
versus large-scale farming. Large-scale agriculture has gained supremacy
partly through the systematic use of manufactured chemicals; small-scale
agriculture, with larger numbers of people involved, could eliminate or at
least reduce the chemical dependency. The present situation is not at all
healthy; and, with current federal policies that reward bigness, the situation can only get worse. In his famous book, *An Agricultural Testament*,
Sir Albert Howard proposes a sensible remedy:

> The situation can only be saved by the community as a whole. The
> first step is to convince it of the danger and to show the road out
> of this impasse. The connection which exists between a fertile soil
> and healthy crops, healthy animals and, last but not least, healthy
> human beings must be made known far and wide. As many resi-
> dent communities as possible, with sufficient land of their own to
> produce their vegetables, fruit, milk and milk products, cereals,
> and meat, must be persuaded to feed themselves and to demon-

strate the results of fresh food raised on fertile soil. An important item in education, both in the home and in the school, must be the knowledge of the superiority in taste, quality, and keeping power of food, like vegetables and fruit, grown with humus, over produce raised on artificials.[2]

Howard goes on to say that the processes of decay must balance the processes of growth; it is in the forest, as death and decay restore nutrients to the soil, where the wheel of life can be most clearly observed and appreciated. A successful agriculture imitates nature in this cyclic balancing. But, since the Industrial Revolution, the processes of growth have been artificially speeded up; with the development of chemical fertilizers, agricultural production has increased many times over. Sir Albert contends that a farmer becomes a "bandit" when plants are artificially stimulated to quick and heavy growth: the theft lies in the loss of soil fertility, as natural processes of humus formation are overpowered and neglected.[3] In the context of an undiversified agribusiness, this deterioration of the biological health of the soil grows at an alarming rate.

Nature's response to such imbalance, Howard says, is to accelerate the rate of decay by means of pests, parasites, and disease. But the technician intervenes in this natural corrective too by means of "poison sprays, vaccines and serums and an expensive system of patent medicines, panel doctors, hospitals, and so forth."[4] The processes of growth hypertrophy while those of decay atrophy, although a rise in the rate of incurable diseases may well be nature's way of forcing a new balance.

Howard did develop an organic method to accelerate decay: the Indore process of composting, worked out in India between 1924 and 1931. This method created humus, in as short a time as ninety days, by the careful composting of all possible plant and animal (including human) waste. This humus, added to the soil and worked in, provides the substances from which healthier and more disease-resistant plants and animals—and, consequently, human beings—might flourish. Sir Albert says that the "slow poisoning of the life of the soil by artificial manures is one of the greatest calamities which has befallen agriculture and mankind. The responsibilities for this disaster must be shared equally by the disciples of Liebig [those who developed and promoted the 'artificial manures'] and by the economic system under which we are living." He predicted that if foods were raised on truly fertile soil, and produce sold fresh, "at least half the illnesses of mankind" would disappear. "In the years to come,"

Sir Albert concluded, "chemical manures will be considered as one of the greatest follies of the industrial epoch."[5] (Masanobu Fukuoka has carried natural farming even further than Albert Howard. In *The One-Straw Revolution*, Fukuoka lists his "four principles": no cultivation, no chemical fertilizer or prepared compost, no weeding by tillage or herbicides, and no dependence on chemicals.[6])

II

One need not be a trained scientist to observe the degree to which the bulk of our food is "scientifically" tampered with. An attentive stroll down the aisles of any large supermarket, stepping of course to the beat of a Muzak jingle, is research enough to demonstrate a basic fact of industrial farming: the food made available to us is overly refined, processed, and preserved by modern industrial techniques whose goals are a striking visual impact on the "consumer," a long "shelf life" for the retailer, and an often stunning profit margin for the processor. (Note that I do not ascribe any of this "stunning profit margin" to the *producer*.)

Enough has been said already—from Adelle Davis to Frances Moore Lappé to Jim Hightower to Lewis Herber—about the decidedly nonnutritional value of much that is passed off to us as "good" food. It is an irony that compels our attention: at the very time when medicine proclaims amazing new competencies in the diagnosis and treatment of disease, society at large is delivered a diet, sanctioned by food corporation nutritionists and the United States Department of Agriculture, that increasingly contributes to poor health. Consider the national scandal of obesity. The retort that "The consumer gets what she demands" is largely bunkum: the advertising "industry" employs highly paid specialists in its amazingly expensive and exacting art of dupesmanship. Perhaps our economists can help us understand what they mean by a "healthy" economy, for such "health" seems similar to the health of junk food and junk bonds. Perhaps they can go on to explain to us why the GNP—which is the "record of all goods and services consumed," including hospital expenses—has been designated the soothsayer, clairvoyant, crystal ball, and fortuneteller of industrial society, and why a constantly growing stock market is somehow evidence of a "healthy" economy. We haven't a clue what a truly ecological economy would look like. It's hard to think about things for which we have no thought. Therefore all "real" policy initiatives are oriented toward extending or patching up the dominant system. Only utopian certitudes are considered worthy of consideration. Eutopian alternatives are, for the most part, seen as too trivial to warrant depth analysis.

But, for instance, what's needed is an adequately "retrogressive" diet; and that is rather easy to describe. It would consist simply of natural food grown in unpoisoned soil—whole grains, fresh fruit, vegetables and nuts, a whole lot less sugar, salt, and meat than we Americans have learned to consume, food without chemical additives and coloring agents, and so forth. All this is really common knowledge, and that implies some rather disturbing things about the power of industrial persuasion through advertising, the absence of readily available alternatives, and that "kooky," niche interests will be met and satisfied by "kooky," niche producers—organic market gardeners, for instance, who believe passionately in good, clean food and who will work very hard for relatively little financial return in order to put their economic lives in line with their ecological convictions. The eutopian model lies in the same moral universe as what Theodore Roszak, back in Chapter 6, called the "anarchist faith," a tradition of earthy groundedness that "embraces communal, handicraft, tribal, guild, and village life-styles as old as the neolithic cultures." It is precisely this tradition that industrial civilization has sought to extirpate.

But in order to get literally to the root of the problem regarding food, it is necessary to examine the way in which modern agribusiness deals with soil. The following passage from Edward Hyams' *Soil and Civilization* provides a comprehensive description:

> Husbandry may develop into what is carelessly called scientific farming, a term commonly used to mean intelligent farming. But *real* scientific farming is usually imposed upon the country by the city, and is not necessarily 'good' farming at all, but merely profitable, for a time, to the farmer or his bank.
>
> Something must be said concerning the confusion over the use of this word *scientific*, in connection with farming.
>
> One of the most significant symptoms of our present unbalanced state of mind in the West, is that the adjective *scientific* has become one of uncritical approbation. True, there has been some reaction away from this state of mind among a few intellectuals, a kind of uneasy, if wholesome, *trahison des clercs*. But we are concerned with the majority of the people for whom *scientific* means *good* and who commonly use certain expressions, signifying approval, which have, in fact, atrocious meanings, e.g., *scientific war-*

fare, scientific crime; or employ the adjective, again as one of praise, in contexts descriptive of socially dubious activities, e.g., *scientific salesmanship, scientific advertising.*

By means of this perverse use of language, the expression *scientific farming* has assumed the significance *good farming.* And so, in certain conditions, it may be. Science in agriculture is good when the approach of scientific specialists to the subject is controlled by an ecologist, or by an ecological point of view; when it is biological rather than mechanical; when the scientist's respect for husbandry is profound; his education humane and philosophical; his methods controlled by empirical trials. Such a scientific agriculture has only come into being in the past two or three decades. But scientific farming until recently has often been, and still often is, very bad farming indeed.

Pure scientific agriculture entails an approach to the soil very different than that of the husbandman. The latter, no doubt unconsciously, is aware that he is a symbiont in an elaborate and delicately balanced union of species . . . The scientist too often approaches the soil in the spirit of an industrialist: here, in this dirty stuff underfoot, is material to be transformed into food by an efficient application of chemistry and mechanics to problems of production. Such a state of mind may, and frequently does, lead to the *consumption* of the soil, and that at a rate vastly increased by scientific efficiency . . .

The agricultural industrialist regards soil as an inexhaustible source of wealth, requiring only sufficiently powerful machines and quick-acting chemicals to extract it. For him a field of wheat is a machine for transforming certain chemicals, which he feeds to the roots, by means of photosynthesis and some later processes, into loaves of bread worth money. Such a point of view cannot arise in the countryside itself: the state of mind from which it derives is one peculiar to highly sophisticated urban communities.[7]

So it seems that something that can legitimately be called an agricultural retrogression is in order, even if such changes would be guided by the clearest insights of ecological science. One of these insights is the need for sustained diversification. It is impossible, as Hyams points out elsewhere in his book, for a "subsistence farmer . . . to practice monoculture, a system pernicious both for the soil itself, and for the farmer's character as a citizen."[8] Masanobu Fukuoka goes so far as to say that natural farming "proceeds from the conviction that if the individual temporarily abandons human will and so allows himself to be guided by nature, nature

responds by providing everything."[9] Larry Korn, in his Introduction to
The One-Straw Revolution, says "Mr. Fukuoka believes that natural farming proceeds from the spiritual health of the individual. He considers the
healing of the land and the purification of the human spirit to be one
process, and he proposes a way of life and a way of farming in which this
process can take place."[10]

In face of the monstrous dangers inherent in the modern world, it
seems both sensible and prudent to begin to work toward the integration
of past and present, nature and culture, so that we might bring to the
future an ecologically balanced relationship between human need and
natural fecundity, that any "modification of the environment" might be a
healing of injuries and not another or a deeper wound. As Edward Hyams
insists, a "community must, like a community of any other creatures, be
founded on the soil."[11] Or, as the Japanese farmer Masanobu Fukuoka puts
it, "There is no other road to peace than for all people to depart from the
castle gate of relative perception, go down into the meadow, and return
to the heart of non-active nature. That is, sharpening the sickle instead of
the sword."[12]

III

Frances Moore Lappé and Joseph Collins, in a book entitled *Food First:
Beyond the Myth of Scarcity*, clarify many misconceptions regarding the
worldwide hunger crisis. In their lengthy book, they document a common
and recurring pattern: in countries where there are thousands or even
millions of hungry people, there is often a net *export* of food. They show
this to be true for many Asian, African, and Latin American countries.
The primary causes for this startling contradiction, they maintain, are the
residual patterns left behind (and still enforced) by the imposed economic
structures of colonialism, now strengthened and perpetuated, even in the
absence of direct colonial control, by the international market system and
world finance institutions. According to Lappé and Collins, there is *not* an
absolute shortage of food. On this point they are emphatic: "*No country in
the world is a hopeless 'basketcase'.*"[13] But their analysis doesn't stop there:

> No, we are not crying "conspiracy!" If these forces were entirely
> conspiratorial, they would be easier to detect and many more
> people would by now have risen up to resist. We are talking about
> something more subtle and insidious: a heritage of a colonial order

in which people with the advantage of considerable power sought their own self-interest, often arrogantly believing they were acting in the interest of the people whose lives they were destroying.[14]

Richard J. Barnet and Ronald E. Muller, in their *Global Reach: A Study of Multinational Corporations*, reveal the existence of "export platforms" for agricultural products in many Latin American countries whose problems with hunger, large populations, and inadequate food distribution are well known, and who can ill afford to provide affluent consumers from the industrial North with fresh fruit and vegetables—and in some instances even meat. According to Barnet and Muller:

> Global companies have used their great levels of power—finance, capital, technology, organizational skills, and mass communications—to create a Global Shopping Center in which the hungry of the world are invited to buy expensive snacks and a Global Factory in which there are fewer and fewer jobs. The World Manager's vision of One World turns out in fact to be two distinct worlds—one featuring rising affluence for a small transnational middle class, and the other escalating misery for the great bulk of the human family. The dictates of profit and the dictates of survival are in clear conflict.[15]

This international food market is the outgrowth of capitalist farming that, as the economic historian R. H. Tawney said in *The Agrarian Problem in the Sixteenth Century*, "supplies the link binding agriculture to the market."[16] And what was taking shape then, five centuries ago in England, is now the dominant market pattern worldwide. In an essay entitled "Ecological Effects of Current Development Processes in Less Developed Countries" (in *Human Ecology and World Development*, edited by Anthony Vann and Paul Rogers), Kenneth A. Dahlberg concerns himself with "the shift in thinking regarding agriculture that has occurred with the so-called green revolution":

> The biological idea behind the development of the new varieties of wheat and rice was that if tropical agriculture was to be improved, seeds specifically adapted to tropical conditions would have to be developed. Unfortunately, this partial ecological insight was vitiated by specialized and technological thinking.
>
> Specialization led to a neglect of any consideration of the social and economic dimensions of peasant farming. Technological thinking led to the typical reaction once better seeds were produced:

that they should be promoted universally, and more importantly, that peasant cultures should change—or be changed—to meet the requirements of this new and 'superior' technology. Peasants are thus expected to change cropping patterns, invest any capital they have in tube wells, fertilizers, and pesticides, and generally shift from local, barter-oriented cultures to larger, market-based systems. In short, they are expected to shift from what in most cases is an ecologically sound traditional agriculture to a form of modern industrial agriculture—with all its attendant ecological costs.[17]

Industrial agribusiness is an inappropriate way to feed a hungry world, to put it in the mildest possible terms. More accurately, agribusiness is a form of rational plunder abstractly justified by industrial economic theory, which, in turn, is a kind of secular theology based on the utopian certitudes of predatory civilization, a kind of moral hubris ethically oblivious to all cultural forms other than those of its own creation. In actual practice, industrial agribusiness throws all possible produce into the money-based market; it seeks to overpower "barter-oriented cultures," as Professor Dahlberg points out.

Wendell Berry, in *The Unsettling of America*, says "The old ideal sought to preserve the farmer on the farm; that was of necessity its first objective."[18] But the imposition of industrial techniques on agriculture invariably results in larger and fewer farms; and these farms, in turn, must shed their traditional features if they are to "stay ahead" in the money economy. To be modern, to live an industrial lifestyle, is to abandon barter and subsistence as practical features of life. As Ivan Illich points out in his book, *Toward a History of Needs*:

> To expand life beyond the radius of tradition without scattering it to the winds of acceleration is a goal that any poor country could achieve within a few years, but it is a goal that will be reached only by those who reject the offer of unchecked industrial development made in the name of an ideology of indefinite energy consumption.[19]

Thinkers like Dahlberg, Berry, and Illich are, at present, hardly more than troubled voices pleading for sanity and honesty in the industrial wasteland. Despite the repeated warnings of independent intellectuals, the relentless drive to control nature and industrialize virtually all human activities brings us closer and closer to the "coming boom"—not the economic boom predicted by the late technocratic futurist Herman Kahn, a boom to lift the entire world into perpetual utopian consumerism, but the big boom of social, political, economic, ecological, and military disaster. Oblivious utopia will, in the end, deliver catastrophic dystopia precisely because of its let-them-eat-cake otherworldly idealism.

Only a growing grassroots movement of people getting in deep touch with nature and the past will enable us to restructure industrial priorities and renew cultural vitality. A culture is strong to the extent that people actually create and actively maintain it. Like all good farming, art, and craft, the best is typically done "by hand." Culture takes time, patience, and commitment. A significant reduction in energy consumption, for example, would give us more time to be peaceful in ourselves, for the pace of our lives would be visibly slowed. And a slowing of our lives would enable a greater involvement in the life of our immediate communities.

Since stable culture is created only slowly, the speeding up of time and the quickening of the industrial economy lead inevitably to a thin and brittle culture. As Masanobu Fukuoka says, "True culture is born within nature, and is simple, humble, and pure. Lacking true culture, humanity will perish."[20] Nothing is more urgent than that we learn quickly to slow down.

ENDNOTES

1. Goodman, "Introduction," viii–ix.
2. Howard, *Agricultural*, 220.
3. Howard, *Agricultural*, 25.
4. Howard, *Agricultural*, 37–38.
5. Howard, *Agricultural*, 220, 224, 38.
6. Fukuoka, *One-Straw*, 33–34.
7. Hyams, *Soil*, 127–28.
8. Hyams, *Soil*, 117.
9. Fukuoka, *One-Straw*, 118.
10. Korn, "Introduction," xxv.
11. Hyams, *Soil*, 133.
12. Fukuoka, *One-Straw*, 176.
13. Lappé and Collins, *Food*, 8.
14. Lappé and Collins, *Food*, 101.
15. Barnet and Muller, *Global*, 184.
16. Tawney, *Agrarian*, 216.
17. Dahlberg, "Ecological," 78–79.
18. Berry, *Unsettling*, 37.
19. Illich, *Toward*, 142.
20. Fukuoka, *One-Straw*, 138.

13

Immeasurable Concepts

REPEATEDLY, I HAVE ALLUDED to an idea that needs further clarification and elaboration—namely, an industrial economy forces the contraction of agricultural populations, rural cultures, and their economies
of subsistence. It is not in the nature of my skills to provide an exacting
empirical proof for this assertion. But, indeed, the contraction of rural
culture is readily observable; one need not be a credentialed specialist
to understand and comment on the conditions of modern life. What's
needed is truthful insight. When Ralph Borsodi, in *This Ugly Civilization*,
referring to the social effect of the factory system in England, beginning
in the latter half of the eighteenth century, said "An essentially agricultural economy with a small admixture of the commercialism fostered by
the merchant guilds was changed violently into an essentially industrial
economy with a small admixture of agriculture," he was not expressing a
whimsical opinion but uttering an obvious historic truth.[1] Or, consider
this remark by Brooks Adams in *The Law of Civilization and Decay*: "The
factory system was the child of the 'industrial revolution,' and until capital
had accumulated in masses capable of giving solidity to large bodies of
labour, manufactures were necessarily carried on by scattered individuals,
who combined a handicraft with agriculture."[2] Adams goes on to quote a
"charming description of Halifax" from the work of Daniel Defoe:

> The nearer we came to Halifax, we found the houses thicker, and
> the villages greater, in every bottom . . . for the land being divided
> into small enclosures, from two acres to six or seven each, seldom
> more, every three or four pieces of land had an house belonging
> to them.
>   In short, after we had mounted the third hill, we found the
> country one continued village, tho' every way mountainous, hardly

an house standing out of a speaking distance from another; and, as the day cleared up, we could see at every house a tenter, and on almost every tenter a piece of cloth, kersie, or shalloon; which are the three articles of this countries labour ...

This place then seems to have been designed by providence for the very purposes to which it is now allotted ... Nor is the industry of the people wanting to second these advantages. Tho' we met few people without doors, yet within we saw the houses full of lusty fellows, some at the dye vat, some at the loom, others dressing the cloths; the women and children carding, or spinning; all employed from the youngest to the oldest; scarce anything above four years old, but its hands were sufficient for its own support. Not a beggar to be seen, nor an idle person, except here and there in an alms-house, built for those that are ancient, and past working. The people in general live long; they enjoy a good air; and under such circumstances hard labour is naturally attended with the blessing of health, if not riches.[3]

This is a writer's sketch, however pastoral, of a functioning folk culture, roughly in the middle of the eighteenth century. By the early nineteenth, such a scene would have been characterized by poverty and near starvation, and we will look at the reasons soon. But let's start more humbly on our theme.

II

Recently at the local dump, I stumbled on a box of discarded books. Among the treasures was *The Elements of Economics*, a 1947 college textbook, written by Professor Lorie Tarshis of Stanford University. It seemed propitious to learn one's economic theory while standing in the midst of so much shameless waste; but the book was long, so I decided to take it home. I found it an interesting work, in an exceedingly dull sort of way. But because it represents the typical outlook in contemporary economic thinking—not that much has changed over the last decades, except the coded jargon has become more complex—the book seems as good as any to examine for its hidden values, if such they can be called. The undertone, despite the monotonous preoccupation with quantity, is evangelical: salvation comes, through numberless economical works, when we finally achieve the zenith of "peak levels" and "total outputs."

Part One, "The Structure of the Economy," has four small chapters. The second chapter, "The Changing Economy," has a few sentences on agriculture:

> One of the most striking shifts has been the decline in the relative importance of agricultural production. In 1870, more than half the people gainfully employed in the United States were engaged in agriculture . . . By 1900, the proportion so employed had fallen to about 35 percent, and by 1930 to about 20 percent. It is fair to conclude that agricultural production has declined in importance, when set against our total output, from something like 50 percent in 1870 to below 20 percent in 1930 . . . Our country, which in 1870 was predominantly agricultural, had by 1930 become specialized in manufacturing, trade, the service industries, and transportation.[4]

These data obviously indicate a very rapid transformation, historically speaking, from an agrarian to an industrial economy. And, if we look at the population statistics from 1790, the data are even more startling: at the time of the first American census, ninety-five percent of the people lived by farming. (We can rightfully assume the great bulk of this was subsistence.) Professor Tarshis goes on:

> In the late eighteenth century the economy was chiefly agricultural, and most of its products were destined for consumption on the farm or for trading within short distances . . . But manufacturing and the service industries, which by 1945 bulked very large in the total, were of little or no importance. Ours was then a subsistence economy; goods were produced chiefly for the use of the producers. Today, on the other hand, goods are produced to be sold on the market, and subsistence production has practically disappeared. *This change by itself is a significant and profound revolution.* (Emphasis added.)[5]

It is regrettable that such statements are not followed by extensive (and sympathetic) analysis, but are only little peak insights all too quickly submerged in mundane "facts."

In his book *The Response to Industrialism: 1885–1914*, Samuel P. Hays offers some broad data on what he calls "the expansion into the countryside" of the new industrial system:

> To the uncritical observer the record of industrialism has been written in the production statistics, the accomplishments of inventor-heroes, and the rising standard of living of the American

> people. Even more significant, however, were the less obvious and
> less concrete changes: the expansion of economic relationships
> from personal contacts within a village community to impersonal
> forces in the nation and the entire world; the standardization of
> life accompanying the standardization of goods and methods
> of production; increasing specialization in occupations with the
> resulting dependence of people upon each other to satisfy their
> wants; a feeling of insecurity as men faced vast and rapidly chang-
> ing economic forces that they could not control; the decline of
> interest in non-material affairs and the rise of the acquisition of
> material wealth as the major goal in life. These intangible innova-
> tions deeply affected the American people; *here lay the real human
> drama of the new age.* (Emphasis added.)[6]

Business, in other words, was pulling people out of their village culture
and rendering them dependent not so much upon each other (as Hays
would have it) as upon the urban market *system* that was beyond their
control.

But back again to Professor Tarshis. Part Two of the textbook,
"The Operations of the Business Firm: Price and Output," constitutes
the bulk of the work—eighteen long chapters. Part Three, "Money and
the Interest Rate," has three chapters; Part Four, "The National Income
and Employment," has fourteen; Part Five, "International Trade, Output,
and Income," also has three. Finally, in the sixth and last section, called
"Interest Groups in the Economy," we find agriculture sandwiched between the other two "interest groups," Labor and Monopoly.

The reader has no doubt tired of this litany of titles; but these very
names, and the tone of analysis that follows these headings, are in themselves evidence of agricultural contraction and the still-expanding dominance of industry. Agriculture *has* become jammed between Labor and
Monopoly: both Labor and Monopoly have wanted cheap food for their
separate but complementary reasons. Cheap food for Monopoly means
lower wages. Cheap food for Labor means wages have more purchasing
power and can be stretched further. Initially, the early industrialists tied
wages to the price of grain; the cheaper the grain, the lower wages could
fall and the higher profits would rise. Workers, in turn, needed to purchase all the necessities of life in the market; the less spent for food, the
more would be available for other necessities—and, increasingly, industrial luxuries. Between these two powerful stones, the whole grain of rural
culture was ground to a fine economistic dust.

As the Editor's Introduction in *The Elements of Economics* puts it, "This book reflects Professor Tarshis's conviction that the major purpose in teaching economics is to enable the student to understand the urgent problems of our national economy and to participate in their solution." This seems a worthy goal. But what are those "urgent problems"? They are, according to the editor, the "over-all problem of full employment and the optimum functioning of the system."[7]

Now the idea of full employment could be very fruitful—if by "full" one meant functional, dignified, unalienated, self-directed, creative, and ecologically sound. But when the term is used in conjunction with the "optimum functioning of the system," one is well advised to withhold consent. For the "optimum functioning of the system" means stuffing only as many warm bodies as are necessary into offices and factories, generating obedient personnel for the system, profit for the upper echelon of ownership and control, and endless gobs of novel industrial commodities for the consumer—who is in turn manipulated by "scientific advertising." ("The average American who watches five hours of television per day," says Jerry Mander, "sees aproximately 21,000 commercials per year."[8]) All this is structured in such a way so as to maximize profits and managerial control, and to minimize any considerations that impinge on profit and efficiency, as narrowly conceived. Used in this way, both terms—full employment and optimum functioning—indicate a completely materialistic disposition that crowds out any consideration of other cultural or environmental values, including alternative definitions of what it means to work. Work itself has less and less to do with inherited craft and real satisfaction; it is compressed into repetitive, "scientific" efficiency, a process that dehumanizes both the work and the people who "perform" it. Make no mistake: there is real "scientific" intent here. Professor Tarshis, writing in the Preface, underscores the point:

> To say that one believes that the scientific method must be used in economic research is now rather like saying that one believes in the good and the beautiful. *It would be very hard to find an economist who would defend the use of non-scientific method.* But the term is interpreted by each economist to cover precisely those practices which he himself finds most genial and helpful. In keeping with my own preferences, *I have tried to avoid introducing concepts which are immeasurable.* (Emphasis added.)[9]

Compare the preceding remarks with these by Jacques Ellul:

> Political economy is no longer a moral science in the traditional
> sense. It has become technique and has entered into a new ethical
> framework ... This represents a decisive step for the creation of
> technique. The technical state of mind is likewise evident in the
> creation of a precise method (which more and more consists in
> the application of mathematics to economics) and in the precise
> delimitation of a sphere of action. In effect, in order for technique
> to exist, method must be applied to a fixed order of phenomena.[10]

What Ellul calls the "fixed order of phenomena" is, in the final analysis, nothing less than nature and culture. Each must be "fixed" in place before technique can take control. (Is this really any different than what Karl Polanyi refers to as the "smashing up of social structures in order to extract the element of labor from them"?) Work is reduced to an efficient, instrumentalist routine, producing standardized commodities for profit. Can anyone really fail to see why economics has been called the "dismal" science? And why science itself, in its hygienic cleanliness, has a rotten stench?

The artificial scenery is peculiar enough here for us to pause a moment and try to take it in. We have in the background a massive industrial economy whose mottoes are Progress and Growth. In fields between the smokestacks are a few modern agribusinessmen, wearing plastic caps with brand-name advertisements for agribusiness chemicals pasted on above the visors, getting a little country air in their air-conditioned tractors. Nearer, in the foreground, is a cluster of dull, architecturally monstrous and malformed buildings that, although they might house anything from a garrisoned army to the headquarters of a soft drink company, really shelter a few highly trained (and extremely well-paid) academics who are engaged in confronting the "urgent problems of our national economy." The flag these scholars work under bears these words: Good, Beautiful, and Scientific Method. The first two words are there as political disguise and metaphysical joke, intended to confound or placate those people with old-fashioned and antiquated sentiments, people who, like E. F. Schumacher, might insist on economics as if people mattered.

Our scholars' real *raison d'etre* is accommodated under a much more austere rubric: *Avoid Concepts Which Are Immeasurable.* Numerical Growth, being passive before the measuring stick, is a key concept among the empirically minded. Growth, along with Progress, which is the momentum of Growth, forms the very basis of the economists' justification. Their inevitable goal, therefore, given their "scientific" reductionism (and the extent to which their very identities are aligned with and integrated into the industrial utopia) is the optimum functioning of the system. Having discarded as "unscientific" any other means of ascertaining value, of determining the real, our economic scholars are left with numbers whose larger and larger sums indicate Growth and whose continually accelerating totals indicate Progress. For all the esoteric calculus involved in "econometrics," this kindergarten philosophy is what it all adds up to.

But let us return to the Interest Groups and pay a wistful visit to our caged Agriculture on display between growling Labor and roaring Monopoly. In this chapter of economic zoology, Professor Tarshis acknowledges that "The low total income earned in agriculture and relatively large number of persons engaged"—remember, this was 1947—"suggests that farm family incomes are low," and that, for example, the "average income per farm family was $1259 in 1935–36 and . . . this was *the lowest figure for any occupational group*. It is obvious that, compared to those of most other occupational groups, farm incomes have been low."[11] (Emphasis added.)

Why is it, compared to other "occupational groups," farmers tend to have the lowest incomes? (In 1970, for example, income from farming made up less than half the total net income of the "average" farm family.) Well, here are

> . . . three reasons why farm incomes are so low. Agriculture is the most nearly competitive industry in the economy. The demand for most agricultural products is relatively inelastic for price; and variable costs in agriculture are low and very flexible, the costs of starting a farm are very small, and furthermore, it is difficult to adjust output quickly.[12]

Let us examine these reasons. First, and to the heart of the matter, agriculture is not a "competitive industry," but we will come back to this point. Second, it may have been true in 1947 that "variable costs in agriculture" were low and the "costs of starting a farm" were small; but neither of these conditions obtains today. The monetary value attached to land has risen astronomically since 1947. Third, why should it be any more difficult to "adjust output quickly" in farming than, say, in automobile construction? General Motors can shut down the assembly lines, close factories, and send workers home. A farmer can dump milk, bury chicks, slaughter pigs, or let wheat rot in the field. Both farmer and industrialist can, narrowly speaking, "adjust output." The point is that General Motors is a monopoly and automobile workers have pensions and lay-off benefits; but farmers struggle on, self-employed (in a wicked sort of commodity piecework) and in chronic debt to the bank. The rich monopolist can relax while demand increases (and prices rise), but the farmer who tries to play the same game goes bankrupt—as many farmers found out during the mid-seventies when they tried to "strike." Farming is neither Monopoly nor Labor; it is biological, not mechanical; and so the organizing principles of businesses and unions are not applicable to agriculture. The tragedy partly lies in calling agriculture an industry—for then the biological basis of farming is simply ignored and stupidity is added to stupidity.

So why are agricultural commodities "relatively inelastic for price"? Why aren't cars or airplanes or air conditioners "inelastic"? This is the classic contradiction. In the first place, the pattern of economic relationship between urban elite and agrarian producers was forged at the earliest stage in civilization, at civilization's founding. The city extracted surplus production (while setting the price paid the producer—if the producer, that is, was other than a slave) but kept the rewards at a level barely above survival. We can see in this relationship the origins of what has come to be called capitalism and imperialism in our own time.

After the decay of the Roman Empire, the balance of power in Western Europe tilted toward the countryside; but it was a countryside in which feudalism, derived from the Roman estate system, was entrenched: the estate slave became the feudal serf. We might say that the upper class waited out the Dark Age in their country villas.

In England, under the pressure brought to bear in Parliament by the new industrialists, the rural estate-owning gentry lost their edge of power by the middle of the nineteenth century, a key feature of which was the importation of cheap grain. (This helps explain why country folk are often described as reactionary, for any shift of power to the city invariably means yet another loss for them and deeper alienation.) But only with the scientific techniques developed in the midst of the Industrial Revolution was the city able to squeeze the agrarian population down to skin and bones while simultaneously compelling production of agricultural commodities to be increased.

This combination of rural depopulation, larger agricultural holdings, and increased farm yields adds up to an unprecedented control over food prices. In order for people to be constantly consuming the commodities coming out of factories, they must have the money with which to purchase those commodities. In order for "consumers" to have the available money with which to buy industrial commodities, food prices must be kept as low as possible, for more money spent on food means less money spent for nonfood commodities. It's as simple as that. It is deliberate and conscious policy at the level of our national government that keeps food, *especially and particularly at the farm level*, "relatively inelastic for price." And we must not lose sight of the fact that food *processors* are quite often food *monopolists*. In the March 1980 issue of *The Progressive*, Daniel Zwerdling, in an article entitled "The Food Monsters," revealed that

> ...in a recent year, corporate directors of Standard Brands, Procter & Gamble, Del Monte, H. J. Heinz, General Mills, Kraft, and Pepsico all sat together on the board of directors of General Motors; that one of the two controlling stockholders of Pepsico was New York's mighty Bankers Trust, which in turn was controlled by corporate directors from General Foods, Heinz, Campbell, Philip Morris, Kraft, and Nabisco; that the other leading stockholder of Pepsico, along with Bankers Trust, was the titan Morgan Guaranty Trust, which was also a leading shareholder of arch-competitor Coca-Cola as well as super-rivals Anheuser-Busch and Philip Morris; that the corporate board of J. P. Morgan was steered by directors from Procter & Gamble, from Campbell, Coca-Cola, and Standard Brands, while Morgan's stocks were controlled in part by Bankers Trust and Citibank; that the corporate directors of Citibank and its parent Citicorp came from Procter & Gamble, Kraft, Pepsico, Beatrice, and Philip Morris; and that Morgan Guaranty in turn was the top controlling stockholder of Bankers Trust and Citicorp, as well as the second-largest controlling stockholder of—we're back where we began—General Motors.
>
> There's no way to understand the full implications of this economic Mobius strip—we can only add it to the other pieces of the food price puzzle. And as the pieces begin to fall into place, they form a rough picture of the food dilemma that Americans confront: No FTC [Federal Trade Commission] investigations will ever break up the conglomerates—the FTC started its anti-trust effort against the cereal manufacturers back in 1968, and Government officials predict a final ruling is still at least five to ten years off. And that affects just one of hundreds of food product industries. No Congress beholden to business interest will pass tough laws striking at conglomerate power: The latest move in Congress, in fact, is to gut the meager powers of the FTC, not to strengthen them. No zealous prosecutors will halt conglomerate manipulations of the

market: When the state of Colorado convicted Beatrice recently of bribing giant retailers to sell its dairy products, the criminal fine—'the biggest in state history,' an assistant attorney general told me—hit Beatrice like a mosquito, stinging it for about fifteen minutes worth of sales.

The pieces added up: No conventional Government actions will make a significant dent in the inexorable inflation and deterioration of the nation's food supply.[13]

Elasticity and profit are controlled at the farm level by Congress and the Department of Agriculture; such policies are openly peddled as "anti-inflationary" or "free market." But, as Daniel Zwerdling lucidly states, monopoly processors of food are under no such constraint or control. In fact, it is they who tend to control Congress and, subsequently, the regulatory agencies of the federal government. Against this entrenched power, the sincere but politically naive protests of farmers have little impact. The ideology of individualism, that still very much permeates the ranks of farmers, is fostered by monopoly in its devious stumping for "free enterprise." Therefore agribusiness clamors for "expanded exports" as a (hopeless) solution to its entrenched problem of low commodity prices. And these expanded exports, produced through a complex system of federal subsidies, often serve to underprice locally grown agricultural commodities in Third World countries, forcing more of those poorer and less-protected farmers off the land. Thus the farm crisis metastasizes.

Some years ago, I was sent some interesting information by the former populist Governor of Minnesota, Elmer Benson. (Governor Benson has since died, speaking out until the end in behalf of farmers and workers.) What Elmer Benson gave me was a copy of an article, "Consumers Should Know," and it listed eight different foods with the price farmers received for their product compared to the price supermarket shoppers paid for the same product. The list included bread, onions, oranges, tomatoes, potatoes, cabbage, carrots, and lettuce. Obviously farmers grow wheat, not bread; but all other produce items had no intrinsic value added by handlers, haulers, or middlemen. According to the article (the data were compiled in Oklahoma and Texas) farmers got 1 1/2¢ per pound for carrots, that then sold for 39¢ per pound in the store. Potatoes got the farmer 2¢ per pound, but cost the shopper 33¢ per pound. Onions brought the farmer a scanty 1/4¢ per pound, but cost 29¢ per pound in the supermarket. And so the list went on. The percentage markup was stupendous. It is literally shocking to see how small a slice of the retail sale price actually goes to those who grow the food. True farming, it seems, is inherently a Third World occupation.[1]

The government controls farm prices in order to ensure cheap and abundant food. Food prices at the farm level can be controlled—and *are* controlled—with impunity by government "inflation fighters." When all is said and done, and all the strike bluster passes over, the farmer is politically impotent. Farmers are continually faced with that motto of agribusiness—"Get Big or Get Out"—from no less than an authority than former Secretary of Agriculture Earl Butz. (What was brutally refreshing about Butz was that he did not hide the industrial ultimatum behind flowery abstractions.) And reductions in dairy price supports, purportedly to reduce surpluses, not only fail to reduce surplus (if there really is any), they only accelerate foreclosures and the further loss of smaller farms. (The newest program—the 40¢ per hundredweight assessment on milk for dairy herd buyout—would be hilarious if it wasn't so obviously cynical and morally sick. The premise seems to be that when the ship is finally sinking, the rats will eat each other to lighten the load, provided the program is explained to them scientifically.)

Let us return, though, to the idea that agriculture is the "most nearly competitive industry in the economy." The key to the industrial view of agriculture lies precisely in this terminology. In the past, when farming was first and foremost a *culture* of modest subsistence, any allusion to "competitive industry" could not have been perceived as anything but absurd. Agriculture was a way of life. True, this agriculture sold or bartered its surplus produce; but the foundation of its existence was household and village self-provisioning. The industrial model, on the other hand, forced a larger and larger portion of local production into the impersonal and anonymous market while creating the conditions that demanded both a greater volume and a more rapid exchange of money. Property taxes required cash. Competition, in the sense of vying for livelihood through the medium of the monetized market, was a relatively inconsequential component of traditional agriculture. By forcing farmers to identify

1. And this is particularly true for the largely Spanish-speaking workers who harvest the bulk of American fruit and vegetables, many of whom are in the United States as "illegals" because "free trade" agreements (like NAFTA) have made their economic situation at home unliveable. The U.S. exports heavily subsidized corn. Mexico exports economically desperate farmers.

themselves as a "special interest" within the industrial model, economists swept aside as atavistic the whole complex nature of agriculture—both its methods and its social patterns—and initiated successful agribusiness-men into the industrial fraternity.

We have already seen how E. F. Schumacher concluded that the "fundamental principles" of agriculture and industry are in opposition. Industry operates on principles of rational organization and calculated resource extraction: what it makes, it makes to sell. It has no other function. Agriculture, on the other hand, might make use of rational organization if it chose, but its guiding ethos is based on the nurture of life—plant, animal, and human. Edward Hyams gives us a sense of how, in the context of mass migration from Europe to America, the psychological transition from cultural sensitivity to organizational rapaciousness took shape:

> The appalling destruction of soils wrought in less than a century in North America was due simply to the fact that it was techni-cally possible. Virgin soil communities were invaded by men from a neighboring high culture, equipped not only with steel, but very soon with machine tools. Moreover, it is important to realize that this destruction was also *psychologically* possible. The natural environment of primitive, native societies is, to some extent, protected against its human members by the mythology and traditions which these men have accumulated in their contacts with trees and animals and herbs, and which each new generation inherits as a lore and a feeling. But the sophisticated intellectuals who are the men of high civilizations, products of their particular 'Socratic revolution,' uninfluenced by Orphic feeling, are not inhibited in their assault upon a soil community. They do not feel themselves to be merely members of it, even though dominant ones, but are outside nature, God's tenants, given a free hand with the landlord's property.[14]

This assault on nature from people who felt themselves *outside* the natural community, outside of nature, was itself a result, at least in part, of the deterioration of peasant culture in Europe, a deterioration that industry and the factory system were rapidly accelerating. (Can't we also conclude that a civilized religion whose God is outside nature helps produce human personalities—not only "sophisticated intellectuals"—who believe themselves, psychologically and spiritually, outside nature, too? And that the factory system, with its even further removal from nature, was an extension of God's Blueprint?) It was not only the environment of "primitive" societies that was protected from calculated human exploitation; traditional agrarian societies too were infused with concern for Earth, even if people in those societies felt pushed by economic pressures to maximize their yields. If economic pressure—population growth, for instance—was only slowly increasing, an agrarian society could learn to modify its farming practices to meet new needs while maintaining the integrity of the soil. (Consider the terrace agriculture of traditional rice farmers in Asia.) But when pressures are suddenly induced from without, pressures that are sufficiently strong to weaken or even destroy the social and cultural fabric, then an opening is made for the systematic importation of exploitative practices.

It is only in this latter context that reference to "competition" has any bearing. Only an agriculture that has first been commercialized and then industrialized can be called competitive. Regarding the small farmer, the situation is as Professor Tarshis describes: "Because of the weakness of his bargaining position, his income tends to be low."[15] What this means is that the small farmer, the social remnant of a cohesive folk culture, has been crushed by the urban-industrial marketplace; it means that laws, government programs, and economic policy generally are created by and designed for those whose interests in the long run are fatal to both the small farmer and to rural culture.

"When income is high," writes Professor Tarshis, "there are good job opportunities in other industries for the surplus farm population, and hence the movement of labor from the farm to the city goes on."[16] Novels have been written about such tragedies; but the Economic Scientist, looking down from the lofty tower of statistics and charts, applauds. Thomas Hardy, in his *Tess of the D'Urbervilles*, published in 1891, describes just such a circumstance where English farmers, agricultural laborers, and artisans were caught in the pump of rural depopulation:

> These families, who had formed the backbone of the village life in the past, who were the depositaries of the village traditions, had to seek refuge in the large centres; the process, humorously designated by statisticians as 'the tendency of the rural population towards the large towns,' being really the tendency of water to flow uphill when forced by machinery.[17]

For economists to imply, as they do in their bland and complaisant prose, that the wrenching of country people out of rural life is a relatively painless process in which one and all (except, perhaps, for a few grumpy and unenlightened traditionalists) are cheerful and appreciative—"the tendency of the rural population towards the large towns"—gives the lie to the purported "objectivity" of economics as an academic discipline. Economics is the secular theology of the industrial ruling class, an obsequious and perverse kind of civilized mythology. As presently construed, economics stands more as metaphysical distraction than affirmation of real human need or cultural groundedness.

To maintain a "scientific" methodology in which the "immeasurable" is systematically eschewed is to rule out completely the complexities of ethics, feelings, communal values, and traditional reserve. A terrible conflict is thereby created between common culture and sophisticated technics, a conflict in which technicians claim the intellectual high ground. (In the university, the humanities were traditionally the highest expression of civilized values. But in technological civilization, the humanities are shunted aside by computer science, engineering, and business administration. Unable to honorably promote "civilized values" in this instrumentalist context, and unwilling to find common cause with either the remnants of folk culture or radical politics, the practitioners of the traditional humanities turn sour and sardonic and cling with dejected pride to their witty "ambience of civility.")

Unchallenged, the technocratic agenda holds sway. In terms of the "farm sector," there is only one way to proceed. "Thus, the best way to keep farm income high," continues our Professor, "is to keep the national income high. Measures to increase the farmers' share in it are desirable, too, but more important, *they only succeed in so far as they injure the well-being of other members of society.*"[18] (Emphasis added.) What that mild and innocent-looking caveat means is this: rural coherence inhibits the concentration of wealth in the urban-industrial "sector" of the economy. Professor Tarshis is surprisingly clear on this point: farmers can increase their share in the national economic pie only by "injuring the well-being" of other members of society. Capitalism is the proper name of any economic system that pulls capital into huge central pools; to facilitate the movement of all forms of capital, the market must be monetized; all factors that impede the concentration of capital must be broken down. Raw materials must be kept as cheap as possible, methods of extraction as rationalized as possible, labor as depressed as possible. All roads must lead to Rome; all capital must converge. Capitalism is civilization in full economic rationality, with its putative metaphysical ideal of aristocratic opulence for all. (That, at least, is its public pretense; its actual functioning is altogether another matter.)

It is possible to see broader implications by changing a few words in the remarks quoted just above: "Thus, the best way to keep Third World income high is to keep First World income high. Measures to increase the underdeveloped share in it are desirable, too, but more important, they only succeed in so far as they injure the well-being of advanced industrial society." This is the point at which capitalism is outright imperialism, explicitly so.[II]

The last (and forty-sixth) chapter, "A Resurvey of the Economy," concludes like this:

> While it may be legitimate for each group to try and improve its lot at the expense of competing groups, the struggle to do so becomes anti-social when it causes *a reduction in the total output* of the economy. The damage done by such a struggle can be most successfully prevented, not by legislative restraint, but by society's adopting measures to keep the total output as high as possible, its composition as nearly ideal as possible, and the distribution as fair as possible. When that is done, the economic problem will be solved. (Emphasis added.)[19]

Here our Professor ends the economic sermon with notes of concern regarding those who "injure the well-being" of others and the "anti-social" effects of those who "try to improve [their] lot at the expense of competing groups." And how are we to identify such undesirable behavior? Not by excessive concentrations of wealth and power in the hands of a few, not by the tyranny of repressive governments, nor by the injuries sustained by Earth—but by noticing when there is a "reduction in the total output of the economy"! The implication in political terms is only too clear: those who dissent from this program are, in that sinister abstraction, undesirably "anti-social."

So we are back, once again, to the Gross National Product as the measure of social health and cultural well-being. Never mind that the fetish of "total output" has destroyed countless communities, ruined untold human lives, despoiled and contaminated vast areas of our only Earth; never mind that this ideology has decimated rural life and destroyed agrarian culture; never mind, even, that our cities are in fearful disarray

---

II. See Edward Said's *Culture and Imperialism* for a deep and penetrating analysis.

from the effects of this brutal and oblivious commercialism. (Is your city in trouble? Try a little slum clearance, and put in a Convention Center.)

No, we are supposed to fold our hands and listen attentively to the professor's pious homily on good works and charitable behavior, according to the gospel of classical economics, with technical glosses, perhaps, by Keynes, Galbraith, Samuelson, and Friedman. Economics is neither disinterested nor objective; if it qualifies as "scientific," then science itself is inherently reductionist and exploitative. Economics has put itself in the service of the technocratic system; it disdains to deal with the culturally "immeasurable." The science of economics has rejected its moral and literary heritage; it has become the slavish hireling of technique. What we need is a revival of immeasurable concepts.

ENDNOTES

1. Borsodi, *This*, 66–67.
2. Adams, *Law*, 256.
3. Adams, *Law*, 256–57.
4. Tarshis, *Elements*, 23.
5. Tarshis, *Elements*, 25–26.
6. Hays, *Response*, 4.
7. Furniss, "Introduction," vii.
8. Mander, *In the Absence*, 79.
9. Tarshis, *Elements*, x.
10. Ellul, *Technological*, 161.
11. Tarshis, *Elements*, 661.
12. Tarshis, *Elements*, 661.
13. Zwerdling, "Food," 26–27.
14. Hyams, *Soil*, 94–95.
15. Tarshis, *Elements*, 665.
16. Tarshis, *Elements*, 667.
17. Hardy, *Tess*, 395.
18. Tarshis, *Elements*, 670.
19. Tarshis, *Elements*, 687.

14

Threshold of Overextension

To suggest that unchecked economic aggression emanating primarily from Western civilization will result in ecological catastrophe or technocratic fascism is still something of an eccentric proposal, even as the evidence for both keeps growing—from global glacier melting to private military contractors. In the words of Joseph Strayer, writing of ancient Rome, the problem of "interesting the inhabitants of the Empire in the fate of their government" is a terribly serious matter, especially in an age of decline and contraction.[1] Most citizens are so addicted to affluence, either directly with things or indirectly through saturation with advertising, that to entertain an alternative of fuller living seems extraordinarily difficult, if not downright loony. People in general are so dulled by excessive organization and bureaucratic dependence that political indifference becomes normative. Add onto what is most certainly a generalized folk disorientation the pervasive effect of television, as it "encourages passivity, isolation, confusion, addiction, and alienation; it homogenizes values and shuts out alternative visions." Jerry Mander goes on to say that television "redesigns us" to be compatible with a centralized, technocratic future.[2]

The past, with its small-scale cooperative self-reliance, has been broken and discredited. A more radical politics based on sharing and environmental limits is mocked as sheer fantasy. Therefore the future seems to depend on continued support for the status quo. What is there to worry about when we live in the greatest country in the history of the world? AM talk radio bends public discontent into hatred for the illegal immigrant, the feminist, the radical Muslim, and the socialist. As both economic and environmental conditions deteriorate in the years ahead, and as military muscle-flexing becomes even more normative as political entertainment, it is likely that those in political control will exact a tighter and tighter ideological conformity on society as a whole. As things fall apart, and as citizenly protest seemingly becomes less effective, the centers of ideological power (and this includes religious power) will attempt to impose an even greater political uniformity. The deeper and more penetrating the critical analysis, the more likely such critiques will be corralled into "free speech zones," specially fenced to keep such analysis contained and humiliated.

On the other hand, if enough people recognize the inherent flaws and limitations within the technocratic agenda and form a coherent political opposition, we could then move with humane steadiness toward cultural renewal. The church, in principle, should be the entity that keeps articulating the truth of our predicament; but it is, in bulk, so locked into mythological abstractions, into creeds, doctrines, and its comfortable otherworldly tax-exempt status, that it has aligned itself overwhelmingly with civilized aggression. Or, where its overt alignment may be lacking or weak, it holds itself aloof from "worldly' or "political" concerns by means of its self-serving teaching on the "two kingdoms"—an otherworldly kingdom for itself and a civil governance kingdom exclusively for the state. It is no easy task to wean ourselves from excessive reliance on bureaucratized technique and commodity distraction, or from the political ideology that rationalizes our greed and gluttony, especially when that rationalization is perversely entangled with religious avoidance. One needs an alternative place to begin. Ivan Illich, in *Toward a History of Needs*, recognized some of these implications years ago:

> For the primitive, the elimination of slavery and drudgery depends on the introduction of appropriate modern technology, and for the rich, the avoidance of an even more horrible degradation depends on the effective recognition of a threshold in energy consumption beyond which technical processes begin to dictate social relations.[3]

In my opinion, American society has already passed the threshold of which Illich speaks. We have already entered the technocratic wilderness. Functional folk culture used to present a cultural alternative, even as it was largely devoid of literary or intellectual support. But self-sustaining folk culture has almost entirely evaporated. The question is whether we can find our way out of our hypercivilization short of absolute catastrophic breakdown, or whether we will, as a political collective, resist and refuse meaningful transition policy until we reach that point where catastrophic breakdown is the only option left.

This is not how things have always been. Throughout history, most human lives were of necessity oriented toward Earth and the local community. The economic goals of life were modestly subsistent. This is true of the earliest gathering and hunting peoples, of the early cultivators, and of peasant societies since the rise of civilization. Civilizations in fact arose on the abundant production of agricultural people. These primary producers were the great base of the civilized pyramid, providing for that most essential of all needs—eating—by their cultivation of the soil. On the toil of these people, and on the fecundity of Earth, urban elites consolidated and expanded their empires. It was the aristocracy within such civilized empires who set a "civilized" standard of consumption far above and beyond subsistence, a pattern uncritically passed along to "revolutionary" democracies determined to be "civilized," a pattern of aristocratic unsustainability that has become "democratized."

In the books of Edward Hyams (*Soil and Civilization*) and Tom Dale and Vernon Gill Carter (*Topsoil and Civilization*), one can find the idea that many early civilizations despoiled and depleted the natural resources indigenous to their locale and, reaching a critical threshold of overextension, entered into unavoidable decline. The authors of the latter book say quite bluntly that "The fundamental cause for the decline of civilization in most areas was deterioration of the natural-resource base on which civilization rested."[4] Forests were cut down, wildlife killed, grasslands overgrazed; erosions set in and soils were depleted. Typically this threshold of destruction was reached as the ruling elites extended their arena of control and their banquet of consumption beyond the range that either the producing people or the fertile Earth could tolerate. It could in fact be said that civilization, especially Western civilization, has yet to learn this lesson and, furthermore, that the power of technology, the magnitude of human population, and the "democratization" of an essentially aristocratic standard of living makes learning the lesson simultaneously more urgent and elusive.

A classic feature of this overextension, that Hyams especially understands, is the consolidation of small farms into huge estates, coupled with a greater and greater orientation toward market production. The foundation of agriculture—subsistent householding with small surpluses—gives way to what we would today call agribusiness, surplus production pure and simple. This process received an entrenched expression under the Roman Empire. Various politicians in the late Republic recognized that the decay of the peasantry was causing Rome to rely on foreign-born mercenaries to fill the ranks of the legions. Despite the grim nature of the concern—war readiness—the insight was clear: the absence of a strong rural culture made the entire society susceptible to deterioration and breakdown. The functional rural linkage that bound society to nature was missing—for the simple reason that this linkage, in the form of peasant culture, had been destroyed. The small farm had been replaced by the aristocratic estate, tended by slaves. The sense of homeland was being reduced to an ironic political abstraction.

A decisive point in the exploitative attitude held by governing elites toward nature and the peasantry was the wedge it drove between people and the proper care of Earth. Compulsory exploitation creates ecological insensitivity. In his book *The Unsettling of America*, Wendell Berry is writing primarily of the modern experience, but his essential message—the question of scale—applies as well to times past:

> The concentration of farmland into larger and larger holdings and fewer and fewer hands—with the consequent increase of overhead, debt, and dependence on machines—is thus a matter of complex significance, and its agricultural significance cannot be disentangled from its cultural significance. It *forces* a profound revolution in the farmer's mind: once his investment in land and machines is large enough, he must forsake the values of husbandry and assume those of finance and technology.[5]

Economic demands thus have deep cultural and ecological implications. When and where those demands are excessive, the net effect is to destroy the "values of husbandry" and hyperactivate the principles of extractive exploitation. Pillage and rape are also susceptible to an imitative trickle-down effect.

Historically speaking, capital concentration has served to promote the high arts of civilization; but this concentration has also served to tax, strain, disrupt, and break down the folk culture of the countryside. Civilizations may glow on the energy extracted from the countryside—both its raw materials and its natural human vitality; but when the energy is depleted, the bright lights go out. Nor is this simply a matter of energy as defined in sheer mechanical terms: the "energy," broadly speaking, that enlivens the high art of civilization derives—or used to derive—from the cultural "organic consciousness" of the common people. This, in turn, made high art rich in complexity, beauty, and depth; it is what gave such enduring power to the music of Bach, for instance, or the painting of Van Gogh. Indeed, the paintings of Van Gogh mark the late nineteenth-century turning point: Van Gogh's friend Paul Gauguin felt artistically compelled to make a decisive move to Tahiti, in order to renew contact with the "primitive." The European "primitive" was disappearing in the Industrial Revolution. (How else are we to understand the relentlessly tragic novels of Thomas Hardy?) By contrast, our present high art tends to reflect raw power, an energy direct from atomic nature and human technical contrivance, an energy of geometry, unformed abstraction, and machines—and, increasingly, a bleak depiction of blasted landscape. There is no longer an indigenous folk art or folk culture from which high art can draw its themes, colors, and inspiration. This helps account for much that is confused and even bizarre in modern civilized art. It is concentrated artistic intellect brooding over a cultural wasteland, over ecological despoliation, and grimly reflecting what it sees and feels. Just as so-called scientific "objectivity" is psychologically derivative from our traditional worship of a God who is outside of and beyond nature, so the artistic "subjective," devoid of a sense of the sacred in nature, suffers from a similar spiritual asphyxiation.

II

The industrializing process "democratized" exploitation while broadening the urban base. Even the working class could now become civilized. A larger and larger portion of people, including those whose parents and grandparents had been peasants, became beneficiaries of global plunder and learned to be proud of their civility. Every man a king, every home a castle.

Modern industrial society displaced its agrarian population more completely than any classical empire, while simultaneously forcing into being an expanded commodity production through the use of complex machinery, manufactured fertilizers, and chemical poisons. In our time, at least four overlapping "revolutions" can be identified: the Industrial, the Educational, the Scientific, and the Managerial. (Jacques Ellul, lumping all these and more under a single heading, would simply say the "Technological Revolution.") In this matrix of highly rationalized progress, the character of traditional rural life has been practically obliterated. Science, as the new purveyor of truth and goodness, has been used to clearcut the thickets of tradition. Rationalized progress hates traditional "superstition"—opaque feelings and emotional considerations that cannot tolerate the glare of "pure" knowledge—the way the inquisitorial church hated heresy. Indeed, both may be rooted in a merciless quest for nonexistent Absolute Truth, purely abstract ideals. This is the kind of slippery literalistic thinking that strives to actualize its glittering ideals by strictly rational means. Technocratic thinking is a close cousin of religious fundamentalism. Science, in some ways, is an attitudinal extension of religious inquisition.

"Superstition" can mean any sentiment or attraction, any belief or conviction, that does not accept the orthodox banning of all concepts that are immeasurable. So any "unfounded" hesitation a farmer might have, say, about applying herbicides or pesticides to fields that have never been poisoned, must be overcome by the superior, self-assured, fact-filled "truth" of science. And since quantity is the most hospitable of all possible measurables, a few well-placed comments on the probability of increased yields may be enough to overcome the farmer's resistance or indecision. In an industrial economy that squeezes out all but the most productive and "efficient" farmers, many sensitive farmers sadly conclude that they must move with the pack of technocratic wolves or perish with the organic lambs.

But if chemical science and technological wizardry now shepherd the nation's agribusinessmen, while whispering in their ears ever more fantastic fantasies of larger holdings and increased yields, these conditions represent the bleak culmination of social and economic convulsions that have their origins in northwest Europe—according to the economic historian R. H. Tawney—back in the sixteenth century. While Tawney acknowledges in his book, *The Agrarian Problem in the Sixteenth Century*, that in England the "statistical evidence reveals no startling disturbance in area enclosed or population displaced," he nevertheless already sees here the gathering of commercial forces—foreign commerce, company promoting, the money market, among other things—that in the following centuries, leading up to the factory revolution, tore the fabric of society, especially agricultural society, apart. "It was the development of the large capitalist farmer," Tawney goes on, "which supplied the link binding agriculture to the market and causing changes in prices to be reflected in changes in the use to which land was put."[6] Another economic historian, Karl Polanyi, writing in *The Great Transformation*, places the origins of European commercialism even earlier:

Commercialization of the soil was only another name for the liqui-
dation of feudalism which started in Western urban centers as well
as in England in the fourteenth century and was concluded some
five hundred years later in the course of the European revolution,
when the remnants of villeinage were abolished.[7]

Passages from Brooks Adams' *The Law of Civilization and Decay* are even
more detailed and graphic. In a chapter entitled "The Eviction of the
Yeoman" we find:

The manor was the social unit, and, as the country was sparsely
settled, waste spaces divided the manors from each other, and
these wastes came to be considered as commons appurtenant to
the domain in which the tenants of the manor had vested rights.
The extent of these rights varied from generation to generation,
but substantially they amounted to a privilege of pasture, fuel, or
the like; aids which, though unimportant to large property owners,
were vital when the margin of income was narrow.

During the old imaginative age, before centralization gathered
headway, little inducement existed to pilfer these domains, since
there was room in plenty, and the population increased slowly, if
at all. The moment the form of competition changed, these conditions were reversed. Precisely when a money rent became a more
potent force than armed men, may be hard to determine, but certainly that time had come when Henry VIII. mounted the throne,
for then capitalistic farming was on the increase, and speculation
in real estate already caused sharp distress. At that time the establishment of a police had destroyed the value of the retainer,
and competitive rents had generally supplanted military tenures.
Instead of tending to subdivide, as in an age of decentralization,
land consolidated in the hands of the economically strong, and
capitalists systematically enlarged their estates by enclosing the
commons, and depriving the yeomen of their immemorial rights.

The sixteenth-century landlords were a type quite distinct
from the ancient feudal gentry. As a class they were gifted with
the economic, and not the martial instinct, and they throve on
competition. Their strength lay in their power of absorbing the
property of their weaker neighbours under the protection of an
overpowering police.

Everything tended to accelerate consolidation, especially the
rise in the value of money. While, even with the debasement of the
coin, the price of cereals did not advance, the growth of manufacturers had caused wool to double in value . . . The conversion of

arable land into pasture land led, of course, to wholesale eviction, and by 1515 the suffering had become so acute that details were given in acts of Parliament. Places where two hundred persons had lived, by growing corn and grain, were left desolate, the houses had decayed, and the churches fallen into ruin . . .

Thus, by degrees, the pressure of intensifying centralization split the old homogenous population of England into classes, graduated according to their economic capacity. Those without the necessary instinct sank into agricultural day labourers, whose lot, on the whole, has probably been somewhat worse than that of ordinary slaves . . .

Before economic competition had divided men into classes according to their financial capacity, all craftsmen possessed capital, as all agriculturists held land. The guild established the craftsman's social status; as a member of a trade corporation he was governed by regulations fixing the number of hands he might employ, the amount of goods he might produce, and the quality of his workmanship; on the other hand, the guild regulated the market, and ensured a demand. Tradesmen, perhaps, did not easily grow rich, but they as seldom became poor.

With centralization life changed. Competition sifted the strong from the weak; the former waxed wealthy, and hired hands at wages, the latter lost all but the ability to labour; and, when the corporate body of producers had thus disintegrated, nothing stood between the common property and the men who controlled the engine of the law.[8]

The effect of this competitive centralization upon cottage handicraft was as ruinous as it was upon agricultural subsistence. Ralph Borsodi says the large-scale production of cheap commodities forced workshop products off the market and drastically reduced domestic production. He claims this change entailed a "disorganization of the economy of the world unprecedented in all history."[9] E. P. Thompson, in *The Making of the English Working Class*, describes in general terms the impact of industrialization on the culture of working people:

Certainly, the unprecedented rate of population growth, and of concentration in industrial areas, would have created major problems in any known society, and most of all in a society whose *rationale* was to be found in profit-seeking and hostility to planning. We should see these as the problems of industrialism, aggravated by the predatory drives of *laissez faire* capitalism. But, however the problems are defined, the definitions are no more than different ways of describing, or interpreting, the same events. And no survey of the industrial heartlands, between 1800 and 1840, can overlook the evidence of visual devastation and deprivation of amenities. The century which rebuilt Bath was not, after all, devoid of aesthetic sensibility nor ignorant of civic responsibility. The first stages of the Industrial Revolution witnessed a decline in both; or, at the very least, a drastic lesson that these values were not to be extended to working people. However appalling the conditions of the poor may have been in large towns before 1750, nevertheless the town in earlier centuries usually embodied some balance between occupations, marketing and manufacture, some sense of variety. The 'Coketowns' were perhaps the first towns of above 10,000 inhabitants ever to be dedicated so single-mindedly to work and to 'fact.'[10]

Thompson's use of the word "fact" brings to mind the diatribe by the industrial schoolmaster Gradgrind in Dickens' *Hard Times*—the bookish pedant who wanted children taught facts, "nothing but facts."[11] But which facts? Only, it seems, instrumentalist facts and moral clichés: two plus two equals four; cleanliness is next to godliness; money breeds and has children.

There is a whole world of alternative "facts" to be learned and explored: the shattered history of our common heritage. It is these facts we need to recover and integrate into our lives. We will never be bold enough to create a liveable future until we are determined to learn for ourselves the truth and meaning of our past. There is no future without a living past.

ENDNOTES

1. Strayer, *Western*, 27.
2. Mander, *In the Absence*, 96.
3. Illich, *Toward*, 114–15.
4. Dale and Carter, *Topsoil*, 20.
5. Berry, *Unsettling*, 45.
6. Tawney, *Agrarian*, 402, 408, 216.
7. Polanyi, *Great*, 179.
8. Adams, *Law*, 200–1, 207, 212.
9. Borsodi, *This*, 66.
10. Thompson, *Making*, 322.
11. Dickens, *Hard*, 1.

15

Ned Ludd and the Grain of Science

Sir Charles P. Snow—later Lord Snow—makes a rather tantaliz-
ing assertion in his famous book, *The Two Cultures and the Scientific
Revolution*. The Industrial Revolution, he says, was "by far the biggest
transformation in society since the discovery of agriculture."[1] Such an exceptionally provocative statement might cause one to expect Lord Snow
to deal rather carefully with those thoughtful people who tried to grapple
with the cultural chaos generated by that "revolution." But Snow manages,
however, to lump "Ruskin and William Morris and Thoreau and Emerson
and Lawrence" in one squirming mass and dispose of their various writ-
ings and practical proposals by saying that their collective contributions
"were not in effect more than screams of horror." Nor does Snow stop
there. He also says that "Intellectuals, in particular, literary intellectuals,
are natural Luddites."[2]

The name Luddite is so heavily loaded with negative connotations
that it is well worth tending to its history. The name comes, of course, from
those groups of English workmen who, in the early nineteenth century,
responded to the assault on their livelihoods by destroying power looms,
shearing frames, and framework knitting machines in the hated factories.
The name Luddite is supposed to derive from a Ned Ludd who allegedly
smashed stocking frames around 1780. As E. P. Thompson says, "Luddism
lingers in the popular mind as an uncouth, spontaneous affair of illiterate
handworkers, blindly resisting machinery."[3] But Thompson also says that
the "conventional picture" of Luddism is not tenable:

> What was at issue was the 'freedom' of the capitalist to destroy the
> customs of the trade, whether by new machinery, by the factory-
> system, or by unrestricted competition, beating-down wages, un-
> dercutting his rivals, and undermining standards of craftsmanship.
> We are so accustomed to the notion that it was both inevitable

and 'progressive' that trade should have been freed in the early 19th century from 'restrictive practices,' that it requires an effort of imagination to understand that the 'free' factory-owner or larger hosier or cotton-manufacturer, who built his fortune by these means, was regarded not only with jealousy but as a man engaging in *immoral* and *illegal* practices. The tradition of the just price and the fair wage lived longer among 'the lower orders' than is sometimes supposed. They saw *laissez faire*, not as freedom, but as 'foul Imposition.' They could see no 'natural law' by which one man, or a few men, could engage in practices which brought manifest injury to their fellows . . .

[W]e may see Luddism as a moment of *transitional* conflict. On the one hand, it looked backward to old customs and paternalist legislation which could never be revived; on the other, it tried to revive ancient rights in order to establish new precedents. At different times their demands included a legal minimum wage; the control of the 'sweating' of women or juveniles; arbitration; the engagement by the masters to find work for skilled men made redundant by machinery; the prohibition of shoddy work; the right to open trade union combination. All these demands looked forwards, as much as backwards; and they contained within them a shadowy image, not so much of a paternalist, but of a democratic community, in which industrial growth should be regulated according to ethical priorities and the pursuit of profit be subordinated to human needs.[4]

By these standards, this present essay upholds Luddism as an honored antecedent.

Now for a man who claims to be writing a book whose purpose is to bring the "literary culture" and the "scientific culture" closer, Lord Snow can hardly assert that his words serve the lofty purpose of cultural reconciliation. The effect of his writing on the imagination is rather like finding gasoline in the pail of a fireman headed toward a burning house.

But Snow says "There is a moral component right in the grain of science itself."[5] Now one would think that Lord Snow, himself a "first-rate" scientist, would take the time and trouble to point out what that "component" is, where it comes from, and what restraints it imposes on the procedures of the scientific process. But he doesn't bother with such trivialities. In a footnote on the very page in which the above passage appears, Snow approvingly gives examples of what terms like "Subjective," "Objective," and "Philosophy" have come to mean in "contemporary technological jargon." The word Philosophy, he says, "means, 'general intellectual approach or attitude' (for example, a scientist's 'philosophy of guided weapons' might lead him to propose certain kinds of 'objective research')." One is left, rather bewildered, still looking for the indigenous moral component in the grain of science. (Perhaps C. P. Snow should have had a chat with Solly Zuckerman on the subject of nuclear illusion and reality.)

Snow also seems to feel that nothing of real enduring value was lost to human culture in the carnage of the Industrial Revolution—at least nothing of any sustaining merit. His view of the past is immortalized in that marvelously prejudiced Hobbesian idea (which he quotes) that life, for the majority of humankind, has always been "nasty, brutish, and short." His opinion of what is to be done to correct this evolutionary backwardness winds down to a very stark proposition: "Industrialization is the only hope of the poor."[6]

II

The art historian Arnold Hauser, whose four-volume *Social History of Art* stands as a classic in its field, deals at considerable length with the two famous nineteenth-century Englishmen whom Lord Snow openly disdains—John Ruskin and William Morris. Ruskin was an internationally known critic of both art and architecture. In his view, the ugliness of modern art and architecture merely reflected the ugliness of industry and the essential unsoundness of commercialism as a philosophy of life. Morris, following Ruskin, was a painter, poet, craftsman, novelist, and pamphleteer of boundless energy whose sensitivity to the squalidness of factory life led him to socialism. But Morris's socialism was decentralist; and the clue to his thought was his constant insistence on the need for quality work and fine craftsmanship.

Like C. P. Snow, Arnold Hauser turns his analysis of modern life on the issue of mechanization. In the fourth volume of his *Social History*, Hauser says that while Morris shows that "art arises from work" and from "practical craftsmanship," he nevertheless "fails to recognize the significance of the most important and most practical modern means of production—the machine."[7] Hauser goes on to say, approvingly, that Ruskin attributed the decay of art to the decay of craft; that with its assembly lines, long hours, and strict division of labor, the factory obstructed "a genuine relationship between the worker and his work," estranged "the producer from the product of his hands," and crushed the "spiritual element."[8]

But in the very next sentence, Hauser insists that in Ruskin's hands the "fight against industrialism lost the barb directed against the proletarianization of the masses."[9] Now this is a strange idea, for it was precisely the factory mode of production that Ruskin opposed; and it was this very factory system, following on the heels of the large-scale enclosures in the latter years of the eighteenth century and the first half of the nineteenth, that *institutionalized* the estrangement of the producer from the product of his hands and prevented a genuinely creative relationship from emerging between the worker and his work. To oppose the factory in which the workers had become mere personnel was to place oneself exactly against the "proletarianization of the masses," if for no other reason than that the factory was precisely the place where the "masses" were proletarianized.

Hauser concluded that particular paragraph by saying that Ruskin was substantially correct in his insistence that "man did in fact lose control over the machine." But Hauser also says that Ruskin "forgot that there was no other way to control the machine than to accept it and conquer it spiritually." It is, however, hard to understand what Hauser intends by the word "spiritually," for on the following page, in a convoluted attempt to rationalize this confusion, he says: "Even the human voice—and even the vocal apparatus of a Caruso—is a material instrument, not a spiritual reality."[10] What then is the spirituality with which we are to "conquer" the machine? We are never really told. (Hauser's "spirituality" is at least as elusive as C. P. Snow's "moral component" in the grain of science.)

These ideas of Hauser's insinuate a peculiar passivity, a melancholy fatalism. Although Hauser says that the factory system was not to be stopped with "polemical pamphlets and protests" and that it was "extremely childish to try," he fails to explain why we need to "accept the machine" or what it means to conquer the machine "spiritually."[11] (And by "machine," Hauser clearly intends the whole industrial system, not the isolated machine-as-tool.) One is reminded here of the pessimism expressed by Sigmund Freud in his *Civilization and Its Discontents*: civilization badly thwarts the human spirit, but there is nothing to be done about it except learn to accept civilized sublimation with rational resignation.

But is it either right or necessary to "accept" the industrial system? And is it legitimate to believe that one can, by so doing, "conquer it spiritually"? What happens if we bring "the machine" in general down to a very concrete specific: nuclear power reactors. Granted that there are many industrially minded people who, like C. P. Snow, would insist that the danger attributed to nuclear reactors is vastly overrated, and that all we need do, once again, is trust the scientist and technician to solve the various problems—from leakage in temporary storage to "permanent" waste disposal, not to speak of the police-state security with which the facilities are protected. Psychologically and politically, it is the same old ruse whereby the technical elite insist they are the only people with the brains and competence to know what to do. The nineteenth-century industrialist and political economist claimed the same status, the same privilege, for themselves. It wasn't true then and it isn't true now. But there is a difference. Terrible as the living conditions were, for example, among the working poor in early nineteenth-century England—infant mortality, according to E. P. Thompson, reached toward fifty percent—the ecological viability of the entire planet was not yet threatened. We are part of the same technological process, to be sure; but the problems have now "matured" and the dangers are much more comprehensive and intense. The margin of subsistence is much curtailed. The commodity system is so encompassing. There is, in the context of so much danger, so little room for error, so little ecological capacity for the perpetuation of our civilized obsessions.

The issue—unchecked technological development—demands our attention. Arnold Hauser, like so many sensitive and intelligent people, comes up to the issue and shies away: he cannot endorse the social conditions that prevailed prior to industrialization, but he can see no way to reach a more natural and more naturally fulfilling society other than by allowing ourselves to be swallowed by the mechanical Leviathan. He embraces industrialization as if it were the will of God. So now we are in the belly of this Leviathan, and the speed with which the mechanical monster is poisoning Earth is purely astonishing. And, in this context, our ability to grapple with the situation is vitiated by our induced passivity. Stuart Ewen, in his *Captains of Consciousness: Advertising and the Social Roots of the Consumer Culture*, gives us a glimpse into how a "philosophy of futility" was deliberately shaped by the advertising media:

> Just as the factory was eradicated from the affirmative vision of productivity, so too was the propagation of a utopian passivity an attempt to neutralize the frustrated passivity of daily life that, even in the admissions of businessmen, increasingly characterized industrial society. Industrial growth in America had institution-

alized monotony as a feature of work and 'disappointment with achievements' as a common malaise, noted business economist Paul Nystrom. It was the absence of any forceful social bonds and the development of a widespread *'philosophy of futility,'* he continued, that might be effectively mobilized in the stimulation of consumption. Speaking of the seeming purposelessness of American industrial life itself, Nystrom noted that 'this lack of purpose in life has an effect on consumption similar to that of having a narrow life interest, that is, in concentrating human attention on the more superficial things that comprise much of *fashionable* consumption.' The mass-produced goods of the marketplace were conceived of as providing an ideology of 'change' neutralized to the extent that it would be unable to effect significant alteration in the relationship between individuals and the corporate structure. 'Fatigue' with the futility of modern life might, if all other avenues of change are eradicated, be channeled toward a 'fatigue . . . with apparel and goods used in one's immediate surroundings.'

The conception of consumption as an alternative to other modes of change proliferates within business literature of the twenties. Given the recent history of anti-capitalist sentiments and actions among the working class, the unpleasant possibility of 'deeper changes' gave flight to a more pacified notion of social welfare that emanated from consumerization. Recognizing the irreversibility of frustration among those who felt trapped in their surroundings, Helen Woodward spoke frankly of consumption as a sublimation of urges that might be dangerous in other form. Admitting that change would be 'the most beneficent medicine in the world to most people,' Woodward offered mass consumption as a means of acting out such impulses within a socially controllable context. 'To those who cannot change their whole lives or occupations,' she began, 'even a new line in a dress is often a relief. The woman who is tired of her husband or her home or a job feels some lifting of the weight of life from seeing a straight line change into a bouffant, or a gray pass into beige.' The basic issues of industrial capitalism were fractionalized, isolated and reduced to trivialities in her formula. 'Most people,' Woodward declared, 'do not have the courage or the understanding to make deeper changes.'[12]

Ewen goes on to say that "Until we confront the infiltration of the commodity system into the interstices of our lives, social change itself will be but a product of corporate propaganda."[13] But here we come up against the very poison that both induces and contains our passivity. Having been stripped of folk competence and noncivilized consciousness, having been dipped (like so many passive candlewicks) in the hot wax of civilized institutions, we lack folk comprehension precisely at that moment when folk intelligence is most urgently needed. How do we confront the infiltration of the commodity system into the interstices of our lives, with its "institutionalized monotony," when that infiltration has not only been sold to us as something culturally superior but also presented as a liberation and reprieve from the poor, nasty, and brutish imprisonment of folk consciousness?

Already by 1920, Oswald Spengler in his *Man and Technics* was warning that the "mechanization of the world" was leading to a "highly dangerous over-tension." Yet both C. P. Snow and Arnold Hauser remained caught in that peculiar compulsion to see this "dangerous over-tension" through to its ominous conclusion, even though it could very well result in the extinction of human life. Snow's jeer about "screams of horror," and Hauser's condescension regarding the childishness of "polemical pamphlets," are not the kinds of dismissive thinking we so sorely need today.

Stuart Ewen relates an event that puts the dilemma in amazingly sharp profile:

> In an address to the American Association of Advertising Agencies on October 27, 1926, Calvin Coolidge noted that the industry now required 'for its maintenance, investments of great capital, the occupation of large areas of floor space, the employment of an enormous number of people.' The production line had insured the efficient creation of vast quantities of consumer goods; now ad men spoke of their product as 'business insurance' for profitable and efficient distribution of these goods. While line management tended to the process of goods production, social management (advertisers) hoped to make the cultural milieu of capitalism as efficient as line management had made production. Their task was couched in terms of a secular religion for which the advertisers sought adherents. Calvin Coolidge, applauding this new clericism, noted that 'advertising ministers to the spiritual side of trade.'[14]

That Calvin Coolidge, a President of the United States, could have said "advertising ministers to the spiritual side of trade," makes one aware just how urgently we are in need of the monkey-wrenching parables of Ned Ludd.

ENDNOTES

1. Snow, *Two*, 24.
2. Snow, *Two*, 26, 23.
3. Thompson, *Making*, 552.
4. Thompson, *Making*, 549.
5. Snow, *Two*, 14.
6. Snow, *Two*, 44, 27.
7. Hauser, *Social*, 108.
8. Hauser, *Social*, 108.
9. Hauser, *Social,* 108.
10. Hauser, *Social,* 109, 110.
11. Hauser, *Social*, 109.
12. Ewen, *Captains*, 84–86.
13. Ewen, *Captains*, 219.
14. Ewen, *Captains*, 32–33.

16

A Proper Balance

WE SEEM TO HAVE come far afield from rural culture. Or have we? Yes, the issues are various and complex, and to attend only to rural culture is both limiting and narrow. But, no, we must not lose sight of the fact that earthy community is our primary evolutionary heritage. Civilization is only five or six thousand years old, and globalized industrialism only two or three generations. Human life in the countryside, in close proximity to nature, is so deep and ancient its time frame disappears beyond the horizon of our comprehension. The renewal of rural culture is thus a kind of eco-evolutionary litmus test by which to measure our commitment to a "new" and alternative society that no longer scorns rural life or has contempt for the past. If we are serious about wanting a liveable future, we must be prepared to face into the abused past, including looking beyond the glitter of civility to the underlying, brutally repressive inequality on which civility has been constructed—a cruel inequality crowned with an arrogant, narcissistic aristocracy whose patterns of affluent consumption we are coaxed into emulating by "scientific advertising." If we want an ecological society, we shall have to learn to live closely with Earth. If we desire a richer community life, we will begin to share in the most elemental tasks. There will be no liveable future without select retrogressions, retrogressions that resemble our village ancestry even as that ancestry is respectfully transformed beyond all prior confinements.

Our cultural disintegration can be dealt with effectively only as we begin to understand that a liveable, durable, holistic culture must be allied closely with nature and natural processes; we must resist and overcome the industrial and scientific tendency to splinter social "solutions"—and human lives—into numberless abstract divisions and technical specializations. This means, in part, that to recreate a healthy and flourishing agriculture, with a sizeable and stable rural population, we must resolve the problems in the regimented factory as we democratically control the capitalist, commodity-intensive economy. It is not enough to say that the root cause of global distress is economic. Karl Polanyi was very much to the point when he said that a "social calamity is primarily a cultural not an economic phenomenon" and that "disintegration of the cultural environment" is the basic cause of social degradation, even though economic forces are the means whereby the cultural environment is destroyed.[1]

At present, as our sophisticated technology produces a bewildering array of commodities at an increasing pace—a pace that can be continually hastened via the broader application of robotics—the affluent people of the world are incited by "scientific advertising" to quicken their rate of commodity consumption, with little regard for environmental degradation or human oppression. Consumers are to race after cleverly packaged commodities to the "scientific" cheerleading of the advertising "industry," a degraded commercial excitation that saturates human consciousness with images of voyeuristic titillation, seduction that ministers to "the spiritual side of trade." Greed and gluttony hide beneath the cloak of progress and prosperity. This proves that "progress," as the civilized substitute for heaven, is a concept, construct, and doctrine larded with deformed values antithetical to the cultural enfleshment of the kingdom of God, whose core values are sharing and conservation. The practitioners of "advertising psychology" have brazenly manipulated what we might call the collective unconscious of the past and drained it of cultural vitality. This sets the stage for politicians who are packaged as electoral commodities by those with money and the most Machiavellian public relations. As folk memory is lost and as hypnotic consumerism becomes normative, the brand-name corporation, sports team, and military strike force provide the locus of loyalty and excitement. A storm of political hysteria is steadily gathering strength on the national horizon, a political storm created by atomized, heated particles swirling free of embedded cultural or spiritual restraint, and shaped into destructive agitation by professional propagandists practicing demagogic meteorology.

With public consciousness so focused, there is little space or time available for attending to the deep and serious problems of economic structure, industrial scale, ecological wounding, and cultural coherence. It is not, however, "childish" to try and draw attention to these issues, even if those who listen are few, even if the analysis and prescription are not wholly adequate. (Whose are?) Again, in Polanyi's words, "If industrialism is not to extinguish the race, it must be subordinated to the requirements of man's nature."[2] The only word needing modification in that sentence, more than sixty years after it was penned, is *man's* nature. Nearly ninety years ago, in *The Acquisitive Society*, R. H. Tawney stated the issue clearly, in the language of his day:

> It is obvious, indeed, that no change of system or machinery can avert those causes of social *malaise* which consist in the egoism, greed, or quarrelsomeness of human nature. What it can do is to create an environment in which those are not the qualities which are encouraged. It cannot secure that men live up to their principles. What it can do is to establish their social order upon principles to which, if they please, they live up and not live down. It cannot control their actions. It can offer them an end on which to fix their minds. And, as their minds are, so in the long run, and with exceptions, their practical activity will be.[3]

The larger issue here is that a "change of system or machinery" would, at this point, require a transformation of cultural values. To control our use of technology does not mean we must grit our teeth in self-denial; to become less dependent on machinery and economic systems implies, even though we don't immediately see it, an unfolding of cooperative conviviality and communal purpose. We may not have such a glut of stuff in an ecological economy, but our spiritual alertness and cultural vitality would be enormously enhanced.

Men and women will never be able to keep up, either as producers or consumers, with the pace of the machines they have invented. Yet it is in the very nature of machines, and the rationale behind their invention and use, to ease human labor. The machine in and of itself is not at fault. To "conquer" the machine, as Arnold Hauser advocates, is possible only by setting definite and real limits to its applicability and use—and also by replacing the structure of limitless private gain with structures that encourage local democratic direction of economic development, for unless people essentially control the economic decisions in their lives, the culture available to them also will not be their own. Nowhere have I found the issue more lucidly conveyed that in Tawney's *An Acquisitive Society*:

> For it is not private ownership, but private ownership divorced from work, which is corrupting to the principle of industry; and the idea of some socialists that private property and land or capital is necessarily mischievous is a piece of scholastic pedantry as

absurd as that of those conservatives who would invest all property with some kind of mysterious sanctity. It all depends what sort of property it is and for what purpose it is used. Provided that the State retains its eminent domain, and controls alienation ... with sufficient stringency to prevent the creation of a class of functionless property-owners, there is no inconsistency between encouraging simultaneously a multiplication of peasant farmers and small masters who own their own farms or shops, and the abolition of private ownership in those industries, unfortunately to-day the most conspicuous, in which the private owner is an absentee shareholder.

Indeed, the second reform would help the first. In so far as the community tolerates functionless property it makes difficult, if not impossible, the restoration of the small master in agriculture or in industry, who cannot easily hold his own in a world dominated by great estates or capitalist finance. In so far as it abolishes those kinds of property which are merely parasitic, it facilitates the restoration of the small property-owner in those kinds of industry for which small ownership is adapted. A socialistic policy towards the former is not antagonistic to the "distributive state," but, in modern economic conditions, a necessary preliminary to it, and if by "Property" is meant the personal possessions which the word suggests to nine-tenths of the population, the object of socialists is not to undermine property but to protect and increase it. The boundary between large scale and small scale production will always be uncertain and fluctuating, depending, as it does, on technical conditions which cannot be foreseen: a cheapening of electrical power, for example, might result in the decentralization of manufacturers, as steam resulted in their concentration. The fundamental issue, however, is not between different scales of ownership, but between ownership of different kinds, not between the large farmer or master and the small, but between property which is used for work and property which yields income without it ... Once the issue of the character of ownership has been settled, the question of the size of the economic unit can be left to settle itself.

The first step, then, towards the organization of economic life for the performance of function is to abolish those types of private property in return for which no function is performed. The man who lives by owning without working is necessarily supported by the industry of someone else, and is, therefore, too expensive a luxury to be encouraged.[4]

As E. F. Schumacher says in a chapter called "Socialism" in *Small is Beautiful*:

> What is at stake is not economics but culture; not the standard of living but the quality of life. Economics and the standard of living can just as well be looked after by a capitalist system, moderated by a bit of planning and redistributive taxation. But culture and, generally, the quality of life, can now only be debased by such a system.
>
> Socialists should insist on using the nationalized industries not simply to out-capitalize the capitalists—an attempt in which they may or may not succeed—but to evolve a more democratic and dignified humane employment of machinery, and a more intelligent utilization of the fruits of human ingenuity and effort. If they can do that, they have the future in their hands.[5]

To achieve the wise and beautiful ends that Tawney and Schumacher propose depends not only on collective political will—that's obvious—but, more elusively, on what we might call ethical satori, spiritual insight with political consequence. As Jesus pointed out to Nicodemus in the third chapter of John, it's not possible to see the kingdom of God unless one has been "born again," that is, moved beyond private selfishness, strategic interest, and *Realpolitik* to universal compassion, global communality, and ethical servanthood. It is precisely here where Christians and churches, with their endemic doctrinal evasions, obstruct the ethical and ecological unfolding of the kingdom of God and prevent ecological socialism from coming into the flesh and blood of eutopian spiritual politics.

If society is more than an "accessory" to the market, as Karl Polanyi insists, if agriculture is more than industry's feedlot, then we must find real alternatives to the current standard-of-living fantasies generated by the advertising agencies in behalf of corporate profit. No longer held in the rigid grip of the authoritarianism of the past, but also freed from the carnival midway of Western affluence, we shall have to rediscover community and nature almost as children, learning for ourselves and from each other a new conduct of life.

The collective organism of the future—if we are to have a future, and if that future is to be liveable—must be composed of countless "cells" of cooperative community. But any political or religious organization that prohibits the growth of a new people's culture is neither viable nor sustainable. The organizational collective (the government) and the organic cooperative (the culture) must be in some kind of symbiotic balance. Socialism and "anarchism" must accept each other as complementary philosophies of life, as ethical spirituality will be their corpus callosum. As R. H. Tawney was asserting already in 1920, socializing those industries presently controlled by the absentee shareholder—and then pruning such industries to their durable, ecological essentials—is a necessary element, and perhaps even a precondition, for the "restoration of the small master" on the farm and in the shop, though one would hope that ethical and ecological socialization at the level of the large-scale would have its cooperative counterpart in the small-scale, too.

II

It is not terribly helpful to criticize and condemn the existing system without suggesting useful alternatives. For those who are satisfied with the system's assurances, however, no alternative is deemed necessary. A proposal may be "novel" or "interesting" if it is mild enough; but serious proposals will merit heavier descriptions, like "utopian fantasies" or "communist plots." Yet as the foreign policy of the United States is increasingly devoted to promoting and protecting "strategic interests," as international threats and military bullying become more commonplace, as the economy becomes shakier and more erratic, as the environment is blotted by Love Canals, Three Mile Islands, Times Beaches, and mountains of uranium tailings, as unemployment becomes "structural" (despite the attempt to flood the "labor market" with former welfare recipients), as educational regimentation becomes increasingly dysfunctional, and as prisons are the new growth industry, more and more people may become uneasy and decide to consider alternatives. What are some practical proposals?

Somewhat arbitrarily we can say there are four levels to deal with: the personal, the communal, the collective, and the planetary. Different people will, with native inclination, put more energy into one or another of these "levels." But there is urgent need for involvement in each sphere. For some, the need may be for gardening and voluntary simplicity; it may be building a house, tinkering with alternative energy, or learning how to farm. For others, personal energy may be better utilized by local networking: creating food co-ops, community healthcare programs, or—perhaps the most pressing—learning how to live in a functional and productive commune. At the level of the collective, the issues become openly public: learning to confront the industrialization of healthcare, agriculture, recreation, education, and so on. This is the point at which individual persons and cooperative groups must join in common cause; and, if such efforts are to have real and sustained impact on public policy, the need may well be for an alternative political party at the local, state, and federal levels—an intentional public body that integrates into a political whole the rightful demands of organic farmers, minorities, environmentalists, feminists, radical workers, visionary teachers, the poor, sick, aged, and all those who stand for Earth and common culture against the predations of the industrial machine. We are, as a result of commercial culture, all too practiced in powerlessness. At some point the alternative vision must become a social force to be reckoned with; it therefore requires a coherent political expression.[1]

Finally, the present crises culminate in the question of planetary survival. We shall have a "global village" only insofar as Green and Rainbow politics can captivate the public imagination. At its most lucid, ecological socialism finds and sustains the balance point between international

1. There is an obvious contradiction here. On the one hand, the various constituencies just itemized—farmers, feminists, visionary teachers, etc.—are already claimed by the Democratic Party. On the other hand, the Democratic Party, especially at its upper reaches, promotes and defends constant economic growth, the maintenance of the American standard of living, and the perpetuation of capitalist economic institutions. Meanwhile, small political entities like the Greens find it extremely difficult to gather a numerically significant constituency for at least two reasons. First, "third" party candidates in a two-party system are virtually unelectable and so are perceived largely as an exercise in a wasted vote. Second, Greens in particular offer a political platform that *challenges* constant economic growth, the maintenance of the American standard of living, and the continuation of capitalist economic institutions; and it may well be that many citizens decline to vote for Green candidates as much or more because Greens threaten their conventional security.

I write this note in late May of 2008, as gas prices begin to top four dollars per gallon and as the sudden spike in prices, driven, it seems, overwhelmingly by Wall Street speculation in oil futures, threatens to disrupt the entire economy, coming immediately after a burst housing "bubble," as unattached, excess money is thrust by private speculating into multiplying itself to the maximum degree possible. To date, the only proposals floated from national politicians have consisted of a federal gas-tax "holiday" and calls for a temporary halt to government purchases of oil stocks for purposes of a national reserve. (Oh, yes—a Senate committee has publicly scolded oil executives for price gouging and greed.) If there have been proposals for the creation of an extensive public rail system and incentives for household, community, and municipal wind and solar technologies, I confess I have missed hearing about them. Perhaps our boldest politicians whisper such radical thoughts to one another in the secure privacy of their offices, but dare not utter them in public for fear of offending their corporate sponsors.

culture and regional integrity. For international culture to be sustainable, the small-scale and the bioregional must be permitted to thrive. This requires a new form of folk culture that, by means of ecological socialism, has truly broken the power of class and privilege. The implication here is that the sharp cleavage between "folk" and "civilized" can in fact be overcome. It further implies that every person can have available a range of options and a richness of culture that, in the past, was obtainable only by a few—artists and bohemians in particular. What this will mean for music and art, for architecture and gardening, for childrearing and international travel, remains to be seen; but my hunch is that the creative potential is simply staggering.

But, right now, we can begin to look as deeply as we are able into how we ourselves are hooked into the system of exploitative affluence (as its beneficiaries or victims or both) and try to work out in our own lives a creative and ecological response. We can reduce our consumption of superfluities. We can eat more unprocessed, unprepackaged food; we can, if at all possible, move to the country with committed friends and learn to raise some portion of our own food. We can trim the more obvious excesses from our predilection for look-alike houses and begin to build for ourselves. We can refrain from buying synthetic clothes—especially with numbers, slogans, icons, and advertisements strewn across front and back. We can avoid the use of high-energy transportation (like jetliners) that, as Ivan Illich points out, involves us in an "involuntary acceleration of personal rhythms."[6] We can begin to get active in behalf of peace and environmental issues, in food cooperatives, in serious crafts and arts. We can learn to entertain ourselves and talk to each other without being electronically plugged in. We can play our own music.

There will be no larger qualitative change in our society until small gatherings are both stable and wide-spread enough to generate the sense of a real alternative to life in the system. However, important as such small gatherings are as social examples, cultural transformation also requires the ethical realization that political democracy devoid of a fundamental economic equality is only an imitation or ersatz democracy, just as the kingdom of God, if kept in a corner cupboard like Sunday china, suffers spiritual atrophy. We cannot have a quality culture until we get past our destructive infatuation with private gain, celluloid glamour, and otherworldly lust. We can engage the kingdom of God by reaching deeper, simultaneously, into community and nature.

On a larger scale, we can readily identify the conditions for a more dignified and slowly paced life. There would, first of all, be a proper balance between rural and urban populations, with a lively fluidity between the two. Every person should be able to participate in a larger cooperative network, thus having a home and useful activity in more than one locale. The restoration of personal competence in useful and productive craft can be advanced by choosing the right kinds of technology. An appropriate technology may indeed be possible in what Ivan Illich calls a "postindustrial, labor-intensive, low-energy and high equity economy."[7]

If a "growing antagonism between town and countryside is . . . inherent in the very process of modern industrialization," as the historian Maurice Meisner says in *Mao's China*, then we will have a proper balance between industry and agriculture only when we limit the purpose and scale of factory-oriented production, and when we find patterns of decentralized public ownership for utilities and larger industries.[8] In *The Closing Circle*, Barry Commoner offers a helpful list of practical proposals:

> If we are to survive economically as well as biologically, industry, agriculture, and transportation will have to meet the inescapable demands of the ecosystem. This will require the development of major new technologies, including: systems to return sewage and garbage directly to the soil; the replacement of many synthetic materials by natural ones; the reversal of the present trend to retire land from cultivation and to elevate the yield per acre by heavy fertilization; replacement of synthetic pesticides, as rapidly as possible, by biological ones; the discouragement of power-consuming industries; the development of land transport that operates with maximum fuel efficiency at low combustion temperatures and with minimal land use; essentially complete containment and reclamation of wastes from combustion processes, smelting, and chemical operations (smokestacks must become rarities); essentially complete recycling of all reusable metal, glass, and paper products; ecologically sound planning to govern land use including urban areas.[9]

To Commoner's wonderful list I would add a call for reduction in the number of automobiles, airplanes, and motorized toys; the development of a mixed transportation system that would embrace unspeedy railroads, bicycles, solar and wind powered ships; a modest use of trucks, tractors, and busses; a reintroduction of horses and mules for pulling wagons and buggies; and the rediscovery of human feet.

These possibilities are within our reach as never before. Our techno-logical know-how can be used to improve the rail system, sailing ships, bicycles, and buggies. There is a critical and pressing need for formulation and enactment of policies that would bring our culture down to Earth. We cannot *create* culture by these policies; but we can, by their promulgation, help bring about the *conditions* in which culture once again has room to grow. Culture is what people do and how they do it. An ecological culture takes time to mature. We need a reduction in high-energy distraction.

We stand at a critical historic threshold. More of what we already indulge in will lead us into chaos, carnage, and rapid decline. We can no longer ruin Earth and, when the situation is no longer tolerable, pull up stakes and move on. There are no new lands to which we can move. Earth is not only much more densely populated than it was in the past, there is no new land to be claimed. Having "conquered" Mother Earth, and gloated over our manly achievement, we realize that learning to live with her at close range is the only option left. We have, therefore, a qualitatively different situation than has ever obtained in human history. One stage of the human mission has been carried to completion, however brutal its enactment. We cannot ruin our cities, soils, and rural cultures, abandon them, and charge out like conquistadores to find (or steal) replacements. It is impossible to buy back what never was for sale. We *must* conserve what we have, for we have reached the limit of the conquistador impulse; we have reached the end of our ability to tolerate imperial invasiveness and manic technological expansion. Both climate change and the nuclear arsenal tell us we've gone way too far down the path of greed and fear.

We are faced, broadly speaking, with only two alternatives. On the one hand, we can continue to rely on progress and growth; this path will, in my view, result in grave calamity—perhaps in human extinction. The other path is more humble; it involves placing culture before industry, people before technology, ecological coherence before economic manipu-lation. This path requires slowing down, a task we talk wistfully about but only few try seriously. It takes personal determination in a surrogate energy system to counteract the industrial pace. Peaceful entropy must work uphill.

I have called the renewal of human culture a "humble" path, and so it is. But it is a humility that opens to the unfolding mysteries of community and natural living. Human culture can be as humble as the person Jesus— as his generosity, his wisdom, and his courage. Civilization is inclined to be as haughty as Christianity—as self-assured, as vain, and as incapable of historical repentance. To choose culture is an earthy choice. It is a choice for faith over belief, organic consciousness over rational consciousness, the life of the body over the abstractions of the mind. It is a choice that acknowledges mortality and accepts death as an inevitable part of life. It has no need to dream of an abstract immortality because the deeper self knows that any possible life on the other side of death is beyond ego control and lies totally in the nature and capacity of the Creator. To really trust God is to let fear dissolve.

These are, however incrementally, real options. The more polluted our environment becomes, the more we lean toward military "solutions" to popular uprisings and conflicts over "strategic interests," the more deeply into technocratic passivity the population sinks, the more urgent—the more desperate—becomes the choice of open-ended life. And the more heavy becomes the chance of organizational death. That's what will happen if what we really trust is fear.

ENDNOTES

1. Polanyi, *Great*, 157.
2. Polanyi, *Great*, 249.
3. Tawney, *Acquisitive*, 180–81.
4. Tawney, *Acquisitive*, 86–87.
5. Schumacher, *Small*, 260–61.
6. Illich, *Toward*, 129.
7. Illich, *Toward*, 115.
8. Meisner, *Mao's*, 97.
9. Commoner, *Closing*, 283–84.

17

Lost Trail or a New Age

E RIC HOFFER, IN *THE Temper of Our Time*, says that "Up to now in this country we are warned not to waste our time but we are brought up to waste our lives." Despite Hoffer's tendency to use mechanical and instrumentalist terminology, he is in principle correct when he says the "efficiency of a society should be gauged not only by how effectively it utilizes its natural resources but by what it does with its human resources," and "the utilization of natural resources can be deemed efficient only when it serves as a means for the realization of the intellectual, artistic, and manipulative capacities inherent in a population."[1]

To love Earth in a vital way, to cultivate the soil on an organic scale, to have a rich communal life filled with simple pleasures, to build and maintain enduring relationships, and to preserve the natural habitat for coming generations—these are goals that require our careful, sustained, and loving attention. Our future can be extraordinarily rich if we summon the courage to affirm our finest dreams with committed action. The potential for great cultural renewal lies before us, a potential that can only unfold by deep sharing and cooperative effort. But first we must be psychologically ready and spiritually willing to make this step, and that means shaking off both the lethargy and the addictive attachments with which our inner sensibility has been padded.

The sociologist Philip Slater, in *The Pursuit of Loneliness*, has said that "for a satisfying society to exist the recognition that people can and must make demands upon one another must also exist."[2] The task here is to learn to enact our demands upon each other in such a way so as not to destroy the fragile flower of personhood—for as never before in history, the new culture will balance personal autonomy with communal coherence. Fulfillment means personal fulfillment; but personal fulfillment is not possible in any sustained way, nor for any great number of people, unless the larger culture is stable and coherent. We are confronted, as Slater aptly puts it, with the "paradox of trying to build a future that does not always look to the future."[3]

A viable, earthy future depends on how fully we accept the past, with all its shames and glories; on the proper balancing of our psychic and bodily impulses with communal and spiritual restraints; on the honest, healthy acknowledgment of our animal functions, so that we might enrich the soil with our wastes and keep the wheel of life in harmony; and on how we might die with peace and reverence and be able to see, in the closing words of Norman O. Brown's *Life Against Death*, "in the old adversary, a friend."[4]

These are exceedingly rich potentialities—but potentialities we will not reach without cultural transformation and major shifts in self-awareness and spiritual comprehension. As Mary Beard put it years ago:

> Perhaps humanity is finding again a lost trail after untold centuries of wandering through blind alleys, following false guides, and experiencing the tragedies of decimation, ruin and death. Possibly it is returning in its advance to the original trail blazed by primitive women—to prime concern with food, clothing and shelter, health and the arts associated with labor, in short with the care of life, all life—prepared to carry its exploration to romantic lengths and altitudes.[5]

The "prime concern" with "the care of life" is what links women's liberation to ecology; and a women's revolution that would not liberate Earth from the bad habits of "democratic" consumerism and the utopian illusions of "Christian" supremacy is, to say the least, an inadequate revolution.

The women's revolution has the potential to bring the whole human family around full circle, if only we can move beyond our preoccupation with manic economic growth, compulsive consumerism, and righteous "defense," and move on to the essentials of community, ecology, and peace. Feminists need to balance their demands for equality *in* the system with even stronger demands for fundamental change *of* the system. Industrial civilization must contract in order to provide room for new growth in libertarian, communal, and ecological culture. Legal equality—or political freedom, for that matter—must be lived or it will not endure. Formal equality is hollow unless filled with cultural reconciliation and spiritual humility.

Lewis Mumford, in *The Transformations of Man*, has said we now

... stand on the brink of a new age: the age of an open world and of a self capable of playing its part in that larger sphere. An age of renewal, when work and leisure and learning and love will unite to produce a fresh form for every stage of life, and a higher trajectory for life as a whole.[6]

But this "new age" is also a "lost trail" found anew. And it is not only the transformation of *man* that is involved, in a strictly psychological sense, but the transformation of the predatory political and economic structures of civilization, as well as the transformation of *woman* and her control of the domestic household. The roots of these structures lie in the past, in the accrued patterns of interaction that have, for centuries, accumulated between women and men, between peasants and aristocrats, bosses and workers. These patterns have given men excessive power in the market-place and women excessive power in the home. We have looked to the castle, and not to the village, for images of ideal living. To prune these imitative excesses and develop a real sharing and blending of earthier functions is a deep and transformative liberation. Betty Friedan, in the closing chapter of *The Feminine Mystique*, addresses this issue:

[W]hen women do not need to live through their husbands and children, men will not fear the love and strength of women, nor need another's weakness to prove their own masculinity. They can finally see each other as they are. And this may be the next step in human evolution.[7]

The next step in evolution . . . Is this too large a conception—that our global crisis is in fact a turning point in the evolution of human consciousness and culture? Surely extinction must be considered a turning point of sorts. Then why not the transformations that are, undoubtedly, the only alternative to extinction?

We are still, by and large, living within the predatory structures of "the system" and the deformed consciousness it extrudes. It must be said, however, that the system *has* served to lift us out of old gender patterns. There is a great deal of liberated conduct that already goes beyond traditional limitations; but the outward conditions have not yet caught up with the inward change of disposition. The system's structure has served its purpose of spreading civilized belief. Let us now have organic and revolutionary faith. As Mumford says:

> As with the early Christians one must prayerfully watch and wait, making every possible conscious preparation, yet realizing that no cold act of will suffices. When the favorable moment comes and its challenge is accepted, thousands and tens of thousands will spontaneously respond to it, stirred by the sense of fellowship the moment will produce. In that act forces that were neutral or antagonistic to any larger plan or purpose will likewise undergo polarization and become actively helpful. Then a new self will be born.[8]

It might be asserted that Mumford's remark balances on the brink of wishful thinking; but it may be that in our jaded rationality we are predisposed to accept the cynicism of *Realpolitik* over the honest prophecy of an open heart.

Yet, at the present time, the true core of the city and the traditional substance of the countryside have both been overwhelmed by the forces of greed and fear. Can we dare to hope in this dark hour, when the power of disintegration has the apparent upper hand, that the cruelties, illusions, and injustices within the present system will fall aside and open the way for a fresh relationship between women and men? And dare we hope that in this new relationship there will be integration, not only between the sexes and "races," but also between the natural and the cultural, between the countryside and the city?

We are faced with either an accelerating drift toward sterility and chaos or a spiritual metamorphosis in which rationality and intuition, emotion and intellect, the body and the mind, the spirit and physical life, can reach energizing synthesis. If we are capable of coolly discussing probabilities of atomic holocaust and ecological catastrophe, then it should not be indecent to explore the territory of miracle. Let us therefore consider miracles as we simplify our lives in the direction of earthy community and spiritual fulfillment.

II

There *is* a way out of the present global crisis. It involves something bigger than change, at least as we normally use the word. We are talking about transformation.

The frontier as historically understood, the outlet for restless human energy, has now been explored. In the last fifty years we have gone to space. A few elite men with spacecraft have "probed Venus," as their language so tellingly informs us. But shooting men into space in long tubes, models of technological homunculi, is, in the present social context, a desperate attempt to evade the vicissitudes of life on Earth. This is not to say that we should permanently refrain from the exploration of space; but it is a fantasy of the lowest rank to suggest space colonies or rocket shots will ameliorate human difficulties on Earth or provide meaningful distraction for emotional confusion and spiritual emptiness. A person of rage is not fit to be an ambassador; a creature who despoils his habitat is not ready to explore the unknown.

We cannot be far from the hub of our accelerating crises. In the depths of breakdown, the conflict and carnage will quite likely be monstrous. It is hard to see how it could be otherwise, for so little is being done, even by people who should know better, to prepare for an ecological future. Because we have been so thoroughly conditioned by both civilized mythology and Christian theology to locate all ultimate salvation *outside* nature, our wills are too numb, our intellects too inoperative, to fully recognize or appropriately respond to these accelerating and enlarging crises. The flip side of rampant hubris is a stupor of moral catatonia. Habits run deep, but in the impending political and economic earthquakes, our normal world will be radically transformed. Then the mysterious human adventure will take on dimensions that, for untold centuries, have been the passionate yearnings of visionaries, poets, prophets, and dreamers. We will be shaken from our otherworldly somnolence.

The renewal of rural culture, the realization of racial equality and gender reconciliation, the implementation of an environmental ethic, the restoration of smaller cities, the simultaneous growth of personal fulfillment and communal cohesion, the reclamation of the natural—all these are elements of the new age. The most dangerous people of our time are those who remain ideologically contemptuous of nature and human nature, all who find it more appealing to contemplate global death than to join hands in behalf of life. It is a common trait of this cynical superiority to indulge in rapturous fantasies of private salvation while picturing scenes of mass destruction for the wicked and the damned. But as Barry Commoner has stated in *The Closing Circle*, the people of the world "are linked through their separate but interconnected needs to a common fate. The world will survive the environmental crisis as a whole, or not at all."[9]

This crisis almost needs a larger word than "environmental." "Environmental" seems too limited and too abstract. But when we consider that "ecology" and "economy" are both rooted in the Greek *oikos*, and that *oikos* means house, we are back again to the retrogressive household and the "lost trail" of caring for life. "Environ" comes from the Latin *viron*, a circuit, to form a ring around, to surround, encircle, or encompass. We are in the web of life; our lives are in the circuits of community and nature.

III

In an essay entitled "From Domination to Cooperation: Ethical and Economic Motivations toward Sustainable Food Systems," Maynard Kaufman asks us to consider what it means "to live within natural energy flows." Kaufman, who was a professor of religion at Western Michigan University in addition to being an organic dairy farmer and cofounder of the Michigan Land Trustees, also asks what it means "to take [Aldo] Leopold's land ethic seriously and to regard ourselves as members of the biotic community." To take our biotic communality seriously, Kaufman says, "means to regard ourselves as consumer organisms who live *within* the energy system rather than as consumers of the harvest who live *outside* the system."[10]

The energy system that Kaufman refers to is not the flow chart of efficiency experts; it is the actual energy circuit of nature. Kaufman clearly understands both the ecological and the cultural implications of maximizing surpluses to feed human beings who no longer feel themselves part of nature. This extraordinary cleavage from nature is the result of excessive civilization and industrial urbanization, of intellectual and religious abstractedness.

The implications of this civilized urban bias are legion. In the environmental movement, preservationists increasingly embrace elaborate efficiency fixes in order to—theoretically—reduce the emission of pollutants. This is not in all instances an undesirable procedure; but as a dominant expression of contemporary environmental thinking, it contains hidden, unexamined, and dangerous assumptions about the permanence of "the system." Kaufman begins to dig into these assumptions when he says that

> . . . as we evaluate the possibility of an agriculture which can be productive within natural energy flows we must attend not only to the distinction between organic and conventional methods, but also to the distinction between production-oriented and subsistence-oriented agriculture. Household food production is

implied by a biocentric ethic in which humans see themselves as living within an ecosystem, and it is as feasible for us in an industrial society as for people in pre-industrial societies, if we can move beyond the assumptions of the market economy. It may even help us feel some solidarity with the majority of poor people in the world.[11]

In other words, we'll get a liveable world when we live again in nature. To say "again" implies that we once did live in nature. And so we did: this is, simply put, the experience and history of all folk culture. The essential distinction between the formal expropriating organization of civilization and the informal subsistence practices of folk culture lies largely in the ecology of food. To rationalize food production beyond a certain point is to undermine and destroy folk culture; and when folk culture disappears, it's only a matter of time before civilization collapses. If folk culture in the past was "below" civilization, we now need a transformed folk culture that surpasses or rises "above" civilization, a folk culture that selects the best aspects of civilization and incorporates them into the ecological, libertarian collectivity of folk wisdom.

Green politics is the natural politics of rural culture and folk society. As Baker Brownell said, "Life under wholesome conditions has a way of assembling itself in a coherent pattern. It has what may be called organic intelligence." Green politics could be the saving grace of civilization: by instituting radical reforms, we could forestall collapse and the immensity of suffering that will accompany collapse. As a society and as a biotic community, we have the material resources and the technical skill by which to make this transformation smoothly and quickly. What's less clear is whether we have the cultural consciousness, the spiritual integrity, or the political will.

ENDNOTES

1. Hoffer, *Temper*, 30.
2. Slater, *Pursuit*, 149.
3. Slater, *Pursuit*, 142.
4. Brown, *Life*, 322.
5. Beard, *Understanding*, 404–5.
6. Mumford, *Transformations*, 191.
7. Friedan, *Feminine*, 377–78.
8. Mumford, *Transformations*, 189–90.
9. Commoner, *Closing*, 292.
10. Kaufman, "Domination," 77.
11. Kaufman, "Domination," 78.

Bibliography

Achebe, Chinua, *Things Fall Apart*. New York: Anchor Books, 1994.

Adams, Brooks. *The Law of Civilization and Decay*. New York: Vintage Books, 1955.

Albrecht, William A. "Introduction." *Our Synthetic Environment* by Lewis Herber. New York: Knopf, 1962.

Barnet, Richard J. and Ronald E. Muller. *Global Reach: The Power of the Multinational Corporations*. New York: Simon and Schuster, 1974.

Beard, Mary. *On Understanding Women*. New York: Greenwood Press, 1968.

Berry, Wendell. *The Unsettling of America: Culture and Agriculture*. San Francisco: Sierra Club Books, 1977.

Borsodi, Ralph. *This Ugly Civilization*. New York: Simon and Schuster, 1929.

Boulding, Elise. *The Underside of History: A View of Women through Time*. Boulder, CO: Westview Press, 1976.

Brown, Norman O. *Life Against Death: The Psychoanalytical Meaning of History*. Middletown, CT: Wesleyan University Press, 1959.

Brownell, Baker. *The Human Community*. New York: Harper and Brothers, 1950.

Buber, Martin. *Paths in Utopia*. New York: Macmillan, 1950.

Childe, V. Gordon. *What Happened in History*. Baltimore: Penguin Books, 1954.

Clough, Shepard B. *The Rise and Fall of Civilization: An Inquiry into the Relationship between Economic Development and Civilization*. New York: Columbia University Press, 1961.

Commoner, Barry. *The Closing Circle: Nature, Man and Technology*. New York: Alfred A. Knopf, 1971.

Cottrell, Leonard. *The Anvil of Civilization*. New York: Mentor Books, 1957.

Dahlberg, Kenneth A. "Ecological Effects of Current Development Processes in Less Developed Countries." In *Human Ecology and World Development*, edited by Anthony Vann and Paul Rogers. New York: Plenum Press, 1974.

Dale, Tom and Vernon Gill Carter. *Topsoil and Civilization*. Norman, OK: University of Oklahoma Press, 1955.

Dewey, John. *The School and Society*. Chicago: University of Chicago Press, 1953.

Dickens, Charles, *Hard Times*. London: J .M. Dent & Sons, 1974.

Durant, Will. *The Reformation: A History of European Civilization from Wyclif to Calvin: 1300-1564*. New York: Simon and Schuster, 1957.

Eliade, Mircea. *The Sacred and the Profane: The Nature of Religion*. Translated from the French by Willard R. Trask. New York: Harcourt Brace & Company, 1987.

Ellul, Jacques. *The Technological Society*. New York: Vintage Books, 1964.

Eriksen, Erik H. *Gandhi's Truth: On the Origins of Militant Nonviolence*. New York: W. W. Norton, 1969.

Ewen, Stuart. *Captains of Consciousness: Advertising and the Social Roots of the Consumer Culture*. New York: McGraw-Hill, 1976.

Freud, Sigmund. *Civilization and Its Discontents*. Translated by James Strachey. New York: W.W. Norton, 1961.

Friedan, Betty. *The Feminine Mystique*. New York: W.W. Norton, 1963.

Fukuoka, Masanobu. *The One-Straw Revolution: An Introduction to Natural Farming*. Emmaus, PA: Rodale Press, 1978.

Furniss, Edgar S. "Editor's Introduction." *The Elements of Economics* by Lorie Tarshis. Boston: Houghton Mifflin, 1947.

Gilk, Paul. *Green Politics Is Eutopian*, Eugene, OR: Wipf and Stock, 2008.

Gilman, Rhoda R., editor. *Ringing in the Wilderness: Selections from the North Country Anvil*. Duluth, MN: Holy Cow! Press, 1996.

Goodman, Paul. *Compulsory Miseducation*. New York: Horizon, 1964.

———. "Introduction." *Living the Good Life* by Helen and Scott Nearing. New York: Galahad-Schocken Books, 1970.

Gorz, André. *Ecology as Politics*. Translated from the French by Patsy Vigderman and Jonathan Cloud. Boston: South End Press, 1980.

Hardy, Thomas. *Tess of the D'Urbervilles*. New York: Dodd, Mead & Company, 1960.

Hauser, Arnold. *The Social History of Art*, Vol. IV. New York: Routledge, 1995.

Hawkins, Gerald S. *Stonehenge Decoded*. New York: Doubleday, 1965.

Hays, Samuel P. *The Response to Industrialism*. Chicago: University of Chicago Press, 1963.

Heilbroner, Robert L. *The Worldly Philosophers*. New York: Simon and Schuster, 1961.

Herber, Lewis. *Our Synthetic Environment*. New York: Knopf, 1962.

Hobsbawm, E. J. *Industry and Empire*. Baltimore: Penguin Books, 1976.

Hoffer, Eric. *The Temper of Our Time*. New York: Harper & Row, 1967.

Howard, Sir Albert. *An Agricultural Testament*. London: Oxford University Press, 1940.

Hyams, Edward. *Soil and Civilization*. London: Thames and Hudson, 1952.

Illich, Ivan. *Toward a History of Needs*. New York: Pantheon Books, 1978.

Jung, Carl. *Memories, Dreams, Reflections*. Recorded and edited by Aniela Jaffé. Translated from the German by Richard and Clara Winston. New York: Vintage Books, 1965.

Kaufman, Maynard. "From Domination to Cooperation: Ethical and Economic Motivation toward Sustainable Food Systems." *Global Perspectives on Agroecology*, edited by Patrick Allen and Debra van Dusen. Santa Cruz, CA: University of California, 1988.

———. "The New Homesteading Movement: From Utopia to Eutopia." *Soundings: An Interdisciplinary Journal*, Vol. LV. (Spring, 1972).

Kinsley, Michael. "Orwell Got It Wrong." *Reader's Digest*, June 1997.

Korn, Larry. "Introduction." *The One-Straw Revolution* by Masanobu Fukuoka. Emmaus, PA: Rodale Press, 1978.

Kropotkin, Peter. *Mutual Aid*. Boston: Extended Horizons Books [no date].

Lappé, Frances Moore and Joseph Collins. *Food First: Beyond the Myth of Scarcity*. Revised and updated. New York: Ballantine Books, 1978.

Leakey, Richard E. *People of the Lake: Mankind and its Beginnings*. Garden City, NY: Anchor Press/Doubleday, 1978.

Lefebvre, Georges. *The Coming of the French Revolution*. Princeton, NJ: Princeton University Press, 1971.

Leopold, Aldo. *A Sand County Almanac*. New York: Sierra Club/Ballantine, 1974.

Mander, Jerry. *In the Absence of the Sacred: The Failure of Technology and the Survival of the Indian Nations.* San Francisco: Sierra Club Books, 1992.

Marcuse, Herbert. *Eros and Civilization: A Philosophical Inquiry into Freud.* Boston: Beacon Press, 1974.

———. *One-Dimensional Man: Studies in the Ideology of Advanced Industrial Society.* Boston: Beacon Press, 1964.

Meisner, Maurice. *Mao's China: A History of the People's Republic.* New York: The Free Press, 1977.

Methvin, Eugene H. "TV Violence: The Shocking New Evidence." *Reader's Digest*, January 1983.

Mishan, E. J. "Ills, Bads, and Disamenities: The Wages of Growth." *The No-Growth Society*, edited by Mancur Olson and Hans H. Landsberg. New York: W. W. Norton & Company, 1973.

Morgan, Elaine. *The Descent of Woman.* New York: Stein and Day, 1972.

Morris, William. *News from Nowhere.* Cambridge, UK: Cambridge University Press, 1995.

Mumford, Lewis. *The City in History: Its Origins, Its Transformations, and Its Prospects.* New York: Harcourt, Brace, and Jovanovich, 1961.

———. *The Myth of the Machine.* New York: Harcourt, Brace and World, 1967.

———. *The Transformations of Man.* New York: Harper Row, 1972.

———. *The Urban Prospect.* New York: Harcourt, Brace and World, 1968.

Norberg-Hodge, Helena. *Ancient Futures: Learning from Ladakh.* San Francisco: Sierra Club Books, 1992.

Polanyi, Karl. *The Great Transformation.* Boston: Beacon Press, 1957.

Pollan, Michael. *In Defense of Food: An Eater's Manifesto.* New York: The Penguin Press, 2008.

Rodale, Robert. "Breaking New Ground." *The Futurist*, February 1983.

Roszak, Theodore. "Introduction." *Small is Beautiful: Economics as if People Mattered* by E. F. Schumacher. New York: Harper & Row, 1973.

Said, Edward. *Culture and Imperialism.* New York: Vintage Books, 1993.

Schumacher, E. F. *Small is Beautiful: Economics as if People Mattered.* New York: Harper & Row, 1973.

Slater, Philip. *The Pursuit of Loneliness: American Culture at the Breaking Point.* Boston: Beacon Press, 1972.

Snow, C. P. *The Two Cultures and the Scientific Revolution.* New York: Cambridge University Press, 1961.

Sotsisowah (John Mohawk). "The Future is the Family." *Akwesasne Notes*, Late Spring 1977 (May issue), Vol. 9, Number 2.

Spengler, Oswald. *Man and Technics: A Contribution to a Philosophy of Life.* Translated from the German by Charles Francis Atkinson. New York: Knopf, 1963.

Strayer, Joseph R. *Western Europe in the Middle Ages.* New York: Appleton-Century-Crofts, Inc., 1955.

Tainter, Joseph A. *The Collapse of Complex Societies.* New York: Cambridge University Press, 1994.

Tarshis, Lorie. *The Elements of Economics.* Boston: Houghton Mifflin, 1947.

Tawney, R. H. *The Acquisitive Society.* New York: Harcourt, Brace & World, 1948.

———. *The Agrarian Problem in the Sixteenth Century.* London: Longmans, Green and Company, 1912.

Thompson, E. P. *The Making of the English Working Class.* New York: Vintage Books, 1966.

Thompson, William I. *The Time Falling Bodies Take to Light: Mythology, Sexuality, and the Origins of Culture*. New York: St. Martin's Press, 1981.

Thoreau, Henry David. *Walden*. New York: Holt, Rinehart and Winston, Inc. [no date].

Toynbee, Arnold. *The Industrial Revolution*. Boston: Beacon Press, 1957.

Waters, Frank. *People of the Valley*. Athens, OH: Swallow Press, 1967.

Yablonsky, Lewis. *Robopaths: People as Machines*. Baltimore: Penguin Books, 1972.

Zinn, Howard. *A People's History of the United States: 1492-Present*. New York: HarperCollins, 1999.

Zuckerman, Solly. *Nuclear Illusion and Reality*. New York: Viking Press, 1982.

Zwerdling, Daniel. "The Food Monsters." *The Progressive*, March 1980.

www.ingramcontent.com/pod-product-compliance
Lightning Source LLC
Chambersburg PA
CBHW061735270326
41928CB00011B/2248